THE GOLDEN AGE OF POLISH PHILOSOPHY

LOGIC, EPISTEMOLOGY, AND THE UNITY OF SCIENCE

VOLUME 16

Logic, Epistemology, and the Unity of Science aims to reconsider the question of the unity of science in light of recent developments in logic. At present, no single logical, semantical or methodological framework dominates the philosophy of science. However, the editors of this series believe that formal techniques like, for example, independence friendly logic, dialogical logics, multimodal logics, game theoretic semantics and linear logics, have the potential to cast new light no basic issues in the discussion of the unity of science.

This series provides a venue where philosophers and logicians can apply specific technical insights to fundamental philosophical problems. While the series is open to a wide variety of perspectives, including the study and analysis of argumentation and the critical discussion of the relationship between logic and the philosophy of science, the aim is to provide an integrated picture of the scientific enterprise in all its diversity.

For other titles published in this series, go to
www.springer.com/series/6936

The Golden Age of Polish Philosophy

Kazimierz Twardowski's Philosophical Legacy

Edited by

Sandra Lapointe

Kansas State University, Manhattan, KS, U.S.A.

Jan Woleński

Jagiellonian University, Cracow, Poland

Mathieu Marion

Université du Québec, Montréal, QC, Canada

and

Wioletta Miskiewicz

IHPST (CNRS/Paris 1/ENS), France and Archiwum Kazimierza Twardowskiego, Poland

Springer

Editors
Prof. Sandra Lapointe
Kansas State University
Department of Philosophy
201 Dickens Hall
Manhattan, KS 66506
USA
lapointe@ksu.edu

Prof. Mathieu Marion
Université du Québec à Montréal
Département de Philosophie
CP 8888, Succursale Centre-Ville
Montréal QC
Canada H3C 3P8
marion.mathieu@uqam.ca

Prof. Jan Woleński
Jagiellonian University of
Krakow
Institute of Philosophy
Grodzka 52
31-044 Krakow
Poland
wolenski@if.uj.edu.pl

Dr. Wioletta Miskiewicz
IHPST (CNRS/Paris 1/ENS)
13 rue du Four
75006 Paris
France
wioletta.miskiewicz@ens.fr

ISBN 978-90-481-2400-8 e-ISBN 978-90-481-2401-5
DOI 10.1007/978-90-481-2401-5
Springer Dordrecht Heidelberg London New York

Library of Congress Control Number: 2009926097

Printed on acid-free paper

Springer is part of Springer Science+Business Media (www.springer.com)

Acknowledgements

This collection of essays proceeds from two conferences that took place within months of each other in the Fall of 2004. The events, which were organised independently, were from their very inception related in a variety of serendipitous ways. For one thing, their respective lists of invited speakers partly overlapped. But most importantly, the organisers shared the same commitment to the history of Polish philosophy and the same conviction as to its contemporary significance.

"Logic, Ontology, Aesthetics. The Golden Age of Polish Philosophy" took place in Montreal on 23–26 September 2004 and was co-hosted by the Université du Québec à Montréal, Concordia University and the Polish General Consulate in Montreal. It gathered some twenty invited speakers and a roughly equivalent number of participants to parallel sessions. Among the speakers were Arianna Betti, Katarzyna Kijania-Placek, Dariusz Lukasiewicz, Bernard Linsky, Wioletta Miskiewicz, Denis Mieville, Claude Panaccio, Douglas Patterson, Roger Pouivet and Jan Wolenski. Their contributions to this volume originate, in part or in whole, from the meeting. For having made the latter possible, the organisers and co-editors of the present book, Sandra Lapointe and Mathieu Marion, would like to thank the Social Sciences and Humanities Research Council of Canada, the Polish Institute of Arts and Sciences, Quebec's Ministry of International Relations as well as the Canada Research Chair in the Philosophy of Logic and Mathematics at the Université du Québec à Montréal.

All other essays, those of Jerzy Bobryk, Anna Jedynak, Grzegorz Malinowski, Paolo Mancosu, Roman Murawski and Urszula Zeglen were originally presented at "Rayonnement de la philosophie polonaise au vingtième siècle. L'héritage philosophique de Kazimierz Twardowski". The event was organised by the Institut d'histoire et de philosophie des sciences et des techniques (IHPST, Paris) and the Husserl Archives (Paris).The organisers and co-editors, Wioletta Miskiewicz and Jan Wolenski, would like to thank the École Normale Supérieure of Paris, the Polish Institute (Paris), the Instytut Adama Miskiewicza (Krakow) and the Polish Ambassy in Paris. For their kind support, the organisers are particulary indebted to Laurence Frabolot, head of ENS's Internationa Office, His Excellency Jan Tombinskii, Polish Ambassador to France and Jacques Dubucs, Head of IHPST.".

As the list of contributors shows, the development of philosophy in Poland is an area of research that has spread to the entire philosophical community, in Europe as well as in America. On the other hand, philosophy in Poland is alive and well and Polish philosophers continue to shape in many ways the contemporary philosophical landscape. We'd like to extend our gratitude to the contributors to this volume, who kindly put up with our editorial demands and patiently suffered the series of revisions their wonderful contributions underwent in the course of the last two years. We would also like to thank Ewa Bolińska and Anna Zielińska for their energetic and effective proof-reading of the Polish, as well as an anonymous referee for many helpful suggestions.

Contents

Contributors

Arianna Betti Vrije Universiteit Amsterdam, Amsterdam, The Netherlands, ariannabetti@gmail.com

Jerzy Bobryk The Institute of Psychology of the Polish Academy of Science and The Institute of Philosophy at the Warsaw School of Social Psychology, Warsaw, Poland, j.k.bobryk@wp.pl

Anna Jedynak Warsaw University, Warsaw, Poland, annjed@wp.pl

Katarzyna Kijania-Placek Jagiellonian University, Cracow, Poland, uzkijani@cyf-kr.edu.pl

Sandra Lapointe Kansas State University, Manhattan, KS, USA, lapointe@ksu.edu

Bernard Linsky University of Alberta, Edmonton, AB, Canada, bernie.linsky@ualberta.ca

Dariusz Łukasiewicz Kazimierz Wielki University, Bydgoszcz, Poland, darek.lukasiewicz@interia.pl

Grzegorz Malinowski Uniwersytet Łódzki, Łodz, Poland, gregmal@uni.lodz.pl

Paolo Mancosu University of California, Berkeley, CA, USA, mancosu@socrates.berkeley.edu

Denis Miéville Université de Neuchâtel (Suisse), Neuchatel, Switzerland, denis.mieville@unine.ch

Wioletta Miskiewicz IHPST (CNRS/Paris1/ENS)/Archiwum Kazimierza Twardowskiego, Warsaw, Poland, wioletta.miskiewicz@ens.fr

Roman Murawski Adam Mickiewicz University, Poznań, Poland, rmur@amu.edu.pl

Claude Panaccio Université du Québec à Montréal, Montreal, QC, Canada, panaccio.claude@uqam.ca

Douglas Patterson Kansas State University, Manhattan, KS, USA, pattersd@ksu.edu

Roger Pouivet Université de Nancy II, Archives Poincaré (CNRS), Nancy, France, roger.pouivet@wanadoo.fr

Jan Woleński Jagiellonian University, Kraków, Poland, wolenski@if.uj.edu.pl

Urszula M. Żegleń Nicolaus Copernicus University, Toruń, Poland, zeglen@uni.torun.pl

Introduction

Jan Woleński and Sandra Lapointe

Polish philosophy goes back to the 13th century, when Witelo, famous for his works in optics and the metaphysics of light, lived and worked in Silesia. Yet, Poland's academic life only really began after the University of Cracow was founded in 1364 – its development was interrupted by the sudden death of King Kazimierz III, but it was re-established in 1400. The main currents of classical scholastic thought like Thomism, Scottism or Ockhamism had been late – about a century – to come to Poland and they had a considerable impact on the budding Polish philosophical scene. The controversy between the *via antiqua* and the *via moderna* was hotly debated.[1] Intellectuals deliberated on the issues of concilliarism (whether the Common Council has priority over the Pope) and curialism (whether the Bishop of Rome has priority over the Common Council). On the whole, the situation had at least two remarkable features. Firstly, Polish philosophy was pluralistic, and remained so, since its very beginning. But it was also eclectic, which might explain why it aimed to a large extent at achieving a compromise between rival views. Secondly, given the shortcomings of the political system of the time as well as external pressure by an increasingly hegemonic Germany, thinkers were very much interested in political matters. Poland was a stronghold of political thought (mostly inclined towards concilliarism) and Polish political thought distinguished itself in Europe

---

J. Woleński (✉)
Jagiellonian University, Kraków, Poland
e-mail: wolenski@if.uj.edu.pl

This introduction offers a very general and equally concise overview of the history of Polish philosophy. More may be found in the following works (in English, French or German): G. Krzywicki-Herburt, "Polish Philosophy", in *The Encyclopedia of Philosophy*, P. Edwards (ed), v. 6, New York, Macmillan, 1967, 363–370; Z. Kuderowicz, *Das philosophische Ideengut Polens*, Bonn, Bouvier, 1988; J. Czerkawski, A. B. Stępień, S. Wielgus, "Poland, Philosophy in", in *Routledge Encyclopedia of Philosophy*, E. Craig (ed), v. 7, London, Routledge, 1998, 483–489; J. Woleński, "Philosophy", in *The Polish Cultural & Scientific Heritage at the Dawn of the Third Millenium*, E. Szczepanik (ed), London, *Polish Cultural Foundation*, 2003, 319–343. Further (selected) bibliographical information about particular periods, movements and persons will be provided throughout the introduction.

[1] On this particular issue Polish philosophers were influenced by Wyclife who had been and still was very popular in neighbouring Prague. Most settled for the modern ideal.

S. Lapointe et al. (eds.), *The Golden Age of Polish Philosophy*, Logic, Epistemology, and the Unity of Science 16, DOI 10.1007/978-90-481-2401-5_1,

at the time by its comparative theoretical superiority. Paweł Włodkowic (known as Paulus Wladimirus), the author of the *Tractatus de potestate papae et imperatoris respectu infidelis* (On The Power Of The Pope And Emperor With Respect To Pagans, 1415) and of the doctrine of *bellum justum* (just war) was the most progressive Polish philosopher of that era. Włodkowic argued that aggression is always unjust and that whenever one seeks to defend one's homeland against external aggressors, all means are justified, including alliances with "non-believers". Włodkowic's argument was a response to recriminations against Poland who had teamed up with the Tartars in the battle of Grunwald (1410). Indeed, as far as non-Christians were concerned Włodkowic was liberal. For instance, one wondered at the time whether it was permissible to use military force to coerce conversions. Włodkowic believed that provided that they respect natural law and live peacefully, non-Christians ought to be tolerated and have a right to enjoy their customs and property. Włodkowic's treatise was presented at the Konstanz Council (1414–1418) and became widely known. It not only inaugurated a glorious political tradition of tolerance in Poland, but also considerably influenced the subsequent development of international law: it is a known fact that the Dutch jurist Hugo Grotius (1483–1645), the co-author of the law of nations, had read *De potestate papae.*

One often thinks of the 16th century as a golden time for Polish culture. The Renaissance made its way from Italy to Poland around 1500 and very soon spread to all aspects of the country's intellectual and artistic life. Poland or, strictly speaking, the Polish-Lithuanian monarchy, was a large and strong country, with considerable agricultural resources. Polish society was amongst the most diverse; its population consisting of Poles, Lithuanians and Ruses (i.e., Ukrainians). Despite the relatively successful attempts by the Jesuit order and the Catholic Church at containing the Reformation – Poland never join the protestant world – religious pluralism was nonetheless an important aspect of its reality. Calvinism was popular among nobility and Lutheranism within urban middle-class. Platonism and stoicism were, in Poland as in other countries where the Renaissance was thriving, very influential – though Aristotelism represented the philosophical mainstream. The former, in particular, substantially informed Copernicus' work, the most remarkable achievement of Polish science at the time. Political philosophy was still thriving. Andrzej Frycz-Modrzewski, for instance, the author *De republica emendanda (On The Improvement Of The Republic*) published in Cracow in 1551 (and in Basilea in 1555), proposed deep political and social reforms – though he defended free will he was very much influenced by Calvin and Erasmus of Rotterdam. In the 16th century, political philosophers in Poland were called upon to solve problems associated with the imminent demise of the Jagiellonian dynasty. In particular, one anticipated that religious pluralism would require scrupulous regulation as well as legal protection. Although Catholicism remained dominant, protestants obtained formal guarantees of religious tolerance and the latter, which was considered to be absolutely essential by most, became a standing political principle. Thomas Jefferson claims that he drafted the American Constitution with the Warsaw Confederation – i.e. the document that ratified the principle of tolerance – in mind.[2] In fact, at the time, Poland was undeniably the most religiously tolerant country in Europe. Many intellectuals oppressed in their own countries emigrated to Poland, among them Faustus Socinus, the Italian antitrinitarist who founded Socinianism, and Amos Komensky (Comenius), a famous Moravian protestant thinker who advocated universal education. This intense cultural life was further stimulated by the rise of new universities such as Vilna, in 1578, as well as the Academies of Zamość and Raków founded by the Polish Brethens in 1594 and 1602, respectively. As a consequence, the average level of education in Poland increased substantially.

The period between 1650–1750 brought a deep political and cultural crisis to Poland. The country was involved in many wars that resulted in economic ruin and in Poland's increasing dependence on Russia. Although the principle of tolerance was not abolished, the Counter-Reformation made some advances and the Polish Brethren, for instance, were forced to leave Poland. They settled in the Netherlands and in England, and influenced philosophical thought in these countries and beyond (e.g., America). We find, for instance, in John Locke's private library some works by the *Fratres Polonorum*, a not so insignificant fact given that Locke's own work was greatly esteemed by the members of the Socinian community. The situation for academic philosophy in Poland was dreadful. Professorial duties at universities and colleges were limited to more or less proficient attempts at a vulgarization of outworn scholastic doctrines. The national system of education was, with only few exceptions, ruled by the Jesuits, and while the latter, among other things, contributed to founding a new university in Lvov in 1652, they did not promote modern ideas. There was virtually no discussion of the great modern philosophical systems or thinkers such as Descartes or Leibniz in Polish institutions of higher education. Even the somewhat refreshing scholasticism of the *scholastici juniori* in Vilna remained indeed scholastic and can hardly be taken to have been a sign of progress. Marcin Śmiglecki did write an extensive textbook of logic that became popular in Oxford: it was reedited 4 times and is evoked by Jonathan Switf in his diaries. Yet, as a whole, Polish philosophy had lost any significant connection with Western thought and became dramatically provincial.

The Enlightenment radically changed this situation. The Polish Enlightenment combined (chronologically) German (Christian Wolff), English (Locke), French (*les philosophes*) and Scottish (Thomas Reid, Dugald Stewart) influences. In its last phase, after 1800 the Polish Enlightenment, were also significantly shaped by Kant's thought. This diversity soon allowed the pluralism that had been characteristic of early Polish philosophy to resurface. French political thought was

[2] The importance of the principle of religious tolerance in Poland is illustrated in the case of the *Fratri Poloni* (Polish Brethens) or Socinians, the most radical wing of Calvinism. While they rejected some of the central dogmas of catholicism, such as the Trinitarian dogma, they nonetheless professed full tolerance and pacifism. See L. Szczucki, Z. Ogonowski, J. Tazbir (eds.), *Socinianism and Its Role in the Culture of XVIth to XVIIth Century*, Warsaw, Polish Scientific Publishers (Państwowe Wydawnictwo Naukowe), 1983.

introduced as an attempt to support Polish independence, which was imperilled by the neighbouring superpowers, Russia, Prussia and Austria. For this reason among others the Polish Enlightenment were concerned with practical matters rather than with theoretical issues. Reform projects were not limited to abstract political questions, but concerned Polish social life as a whole, including education. Teaching at universities and colleges was modernized in order to integrate Lockean-style empiricism, and Étienne de Condillac was commissioned to write a logic textbook by the National Education Committee. Philosophers were mobilised anew: Jan Śniadecki developed a version of positivism inspired by D'Alembert and other Encyclopaedists; Jędrzej Śniadecki (Jan's brother) attempted to reconcile Kant with the Scots; and Józef Kalasanty Szaniawski developed a more or less orthodox version of Kantianism.

The Polish clergy was a remarkably active part of the Enlightenment in Poland. Many of its actors and promoters were indeed clergymen. For this reason, perhaps, the Polish Enlightenment was never entirely hostile to religion. They neither led, as had been the case in France after the Revolution, to institutional secularism nor vindicated atheism – although it must be noted that some representatives of the Church were in fact sympathetic to alternative theologies such as deism. Moreover, the two most important Polish political thinkers at the time, Stanisław Staszic and Hugo Kołłątaj, were both priests. Their views considerably influenced the content of the 3rd May Constitution (1791), the second constitution in History, after the American one.

Unfortunately, political reforms came too late and were useless. Poland lost its independence after three successive partitions, by Russia (in 1773, with Warsaw as the administrative centre), Prussia (1793, where Poznań was the largest city) and Austria (in 1795, where Cracow and Lvov were the main economical and cultural centres). At the beginning of the 19th century, the influence of Romanticism soon led to a fashion of philosophizing that was largely a reaction to the lost independence, *the* national tragedy. The demise was felt all the more keenly after the defeat of the "November Uprising" (against Russia in 1830–1831) and the Tsarist repression that followed. Polish philosophy in the course of this period (and until 1863) can be considered to have been Poland's attempt at a "Nationalphilosophie".[3] The peculiar form it took, what came to be know as Polish "messianism", originated with Józef Hoene-Wroński (also a famous mathematician) and is embodied in the work of a number of great nationalist poets such as Adam Mickiewicz, Juliusz Słowacki, Zygmunt Krasiński. In its substance, Polish messianism is characterised by the belief that Poland and its nation have a unique historical role – although Hoene-Wroński himself pleaded for a kind of Pan-Slavism, with Russia as the leading force. Most Polish Romantics, for instance Józef Gołuchowski, Józef Kremer, Karol Libelt, Bronisław Trentowski or August Cieszkowski (who incidentally coined the term "Historiosophie") considered Poland to be no less than

[3] See A. Walicki, *Philosophy and Romantic Nationalism: The Case of Poland*, Notre Dame University Press, South Bend 1981.

the "Christ" or "Messiah" of Nations: with the suffering of Poles began a New Era that would culminate in the salvation of all people and nations.

The defeat of the January Uprising (1863–1864) led however to radical criticisms of Polish messianic nationalism. The latter was blamed for having promoted unrealistic and even irresponsible political goals not unlikely to have bereft Poland of any chance to regain its autonomy – at least in the near future – as well as to violent acts which caused the death of many. The downfall of Romanticism was ultimately brought about by the "Warsaw (or Polish) positivists", a group of intellectuals that included writers (e.g., Henryk Sienkiewicz, Aleksander Głowacki – also known as Bolesław Prus -, Eliza Orzeszkowa, Maria Konopnicka), journalists (e.g., Aleksander Świętochowski), scientists (e.g. Samuel Dickstein, a mathematician) and philosophers (e.g., Adam Mahrburg). All had a genuine interest in philosophy, although, in most cases, they were also partly autodidacts. Warsaw positivism was influenced by the ideas of August Comte, John Stuart Mill and Herbert Spencer, and later also by Ernst Mach and Richard Avenarius. Even under the unifying aegis of "positivism", Polish philosophy remained pluralist and eclectic, and sought more or less successfully to reconcile points of view that might not have had much in common. For instance, polish positivism accepted (*pace* Comte) that philosophy in order to be scientific must be limited to facts, but admitted psychology among the genuine sciences (*pace* Mill). In social matters, Warsaw positivists insisted that political endeavours be levelheaded and realistic. In particular, they advocated a foundational reform of society, and primarily of education. As they conceived of it, the struggle to accomplish great national feats, independence above all, had to be preceded by a significant rise of the general level of instruction.

The situation of Polish institutions of learning in the 19th century was a difficult one. The University of Vilna was closed by the Tsarist government in 1831, that is, after, the November Uprising. The University of Warsaw which had been established in 1816, was liquidated in the course of the same year. Between 1862–1869 in Warsaw, the "Main School" (Szkoła Główna) served as a replacement for the Polish university; it was a stronghold of Polish positivism. But, in 1870, it was superseded by the Russian Imperial University. After 1815, the University of Lvov became almost exclusively German-speaking and served primarily as a training facility for officials posted in the newly annexed Austrian province of Galicia. Although it was subjected to a series of "germanization" campaigns in 1805–1809 and 1847–1870, the University in Cracow – since 1817 the Jagiellonian University – remained the only academic institution in which the language of instruction was Polish. Poznań (part of the Prussian Dominion), where no academic institution existed, suffered most intellectually.

It should come as no surprise, then, that most Poles studied abroad and that, when they did come home, they brought along the various ideas they had picked up, in Austria, Germany, France, England or Russia. The eventual liberalization of the Austrian-Hungarian Empire in the second half of the 19th century, after the death of Franz II, had a direct impact on Polish culture. Both Galician universities in Cracow and Lvov were repolonized. In 1872, The Polish Academy of Arts and Sciences of Cracow was founded. Warsaw's intellectual life, stimulated in part by the activities of the positivists, and despite the fact that it remained closed until 1918 under Russian rule, was very rich. By contrast, except for the creation of a scientific society in Poznań, nothing new happened in the Polish provinces under Prussian occupation.

In the course of this vibrant period, translations were published whose purpose was to present, in a language adapted for a wider public, Western intellectual achievements. Dickstein founded a journal devoted to mathematics and physics. The rise of *Przegląd Filozoficzny* (*The Philosophical Review*), edited by Władysław Weryho since 1897, provided a remarkable impetus to philosophy. The "Flying (or Floating) University" which had provided clandestine education for women since 1882 and where philosophy was an important part of the *cursus*, saw some 5000 graduates: Maria Skłodowska-Curie was one of them. At the beginning of 20th century, the Warsaw Scientific Society was established and the Flying University was finally granted legal status as the Society for Scientific Courses. Polish academic life thus improved, if not uniformly, at least gradually. Nonetheless, it was only after the Great War, that is, after Poland regained its sovereignty, that a fully integrated system of education and scientific research could be instituted. In 1918, the universities in Warsaw and Vilna almost immediately resumed their activities, and two others – the University of Poznań, the Catholic University of Lublin – were founded.[4]

When Kazimierz Twardowski (1866–1938), a student of Franz Brentano and Robert Zimmermann in Vienna, accepted his nomination to a professorship at the University of Lvov in 1895, his return turned out to be the single most decisive event for the subsequent development of philosophy in Poland. Twardowski had attracted attention as a philosopher in the German-speaking world for his criticism of Brentano and, in particular, for the notorious distinction between the content and object of presentations he put forward in the published version of his Habilitation thesis (1894).[5] From the start, his purpose in accepting the nomination could hardly have been clearer: to introduce scientific philosophy in Poland and – less unassumingly perhaps – to create a school. Twardowski's understanding of the task of philosophy was clearly Brentanian in spirit. Philosophy should rely on scientific methods and resort to clear language. Twardowski, the student, was manifestly closer to present-day analytic philosophy than was the master, Brentano. Twardowski stressed the role of methodological issues and language much more adamantly than Brentano, and his metaphilosophical views systematically favoured investigations in logic and the philosophy of science. In addition, Twardowski was

[4] In fact, the University of Warsaw began its activity in 1915 with the permission of German occupation authorities. Russian Warsaw was taken by the German troops in the very early stage of War World I.

[5] Twardowski's writings are available in English. See K. Twardowski, *On the Content and Object of Presentation*, Martinus Nijhoff, The Hague 1977 (German original appeared in 1894), K. Twardowski, *On Actions, Products and Other Topics in Philosophy*, Rodopi, Amsterdam 1999. See also J. Cavallin, *Content and Object. Husserl, Twardowski and Psychologism*, Kluwer Academic Publishers, Dordrecht 1997.

sceptical of the presumed significance of the reputedly "great" traditional philosophical problems in metaphysics and epistemology. He insisted that philosophers should concern themselves with relevant problems and resist unhealthy speculative inclinations.

Twardowski beginnings in Lvov were not easy. Lvov had no philosophical tradition to speak of and no established teaching model. Hence, at first, Twardowski's courses did not attract many students. But Twardowski was a charismatic teacher and the situation quickly improved. Within a few years he could gather up to 2000 students for his lectures, and up to 200 candidates for his seminars. This was all the more remarkable since he was in fact extremely demanding, not to say plainly brutal. He began his lectures at 7 o'clock in the Spring and Summer, and at 8 o'clock in the Winter. His graduate seminars were organized early in the evening on Saturdays because, he believed, a true philosopher must prefer a seminar over a dance party. In short, his teaching had all the appearances of a military drill. On the other hand, he was available for consultation many hours daily. He also managed to keep an abundant correspondence with his former students. He promoted his first doctors in 1902 and the group of his pupils steadily increased. In 1906, that is, after roughly 10 years of activity, the Lvov *collegium philosophicum* had made its reputation and set itself apart on the Polish intellectual and scientific scene.

Twardowski was the sole force behind this powerful educational thrust. Perhaps precisely for this reason, that is, because he concentrated his energies as an institutional leader, teacher and mentor, Twardowski wrote relatively little. Most notorious is the published version of his Habilitation thesis *On the Content and Object of Representation*. For one thing, it had a tremendous – and generally unacknowledged – influence on the early Husserl and indeed the great merit of having persuaded the latter of Bolzano's importance. His essay "On So-Called Relative Truths" is also famous, among other things, for having instigated a polemic between Leśniewski and Łukasiewicz. This polemic is not unlikely to have led to the latter's logical breakthrough and, thus, to the creation of the first multi-valued logics. Lesser known is his magnificent little essay "On Actions and Products", a piece in which Twardowski analyses the ontological structure of human action and cultural objects. In this volume, two contributions are devoted to the founder of modern Polish philosophy. Jerzy Bobryk's essay offers here a presentation of Twardowski's analyses of human behaviour which will hopefully incite contemporary philosophers of mind and metaphysicians to look upon Twardowski not only as a remarkable pioneer of their field, but as a genuine source of inspiration for contemporary accounts of social and cultural entities. Dariusz Łukasiewicz, on his part, devotes his study to the irreducibly Brentanian themes that emerged through Twardowski's teaching and which shaped in no insignificant way his students' approach to philosophical problems. He identifies two leitmotivs: the "scientific" style of philosophizing and the rejection of Kantianism. Twardowski was not an orthodox Brentanian – after all, *Content an Object* was mostly meant as a criticism of the latter – and while Polish philosophy could hardly repudiate its Brentanian upbringing, this upbringing was not mono-parental. It would be hard to imagine what Polish philosophy would have been without Twardowski's own input.

Twardowski had his own somewhat idiosyncratic vision of Polish philosophy and, in particular, of its overall place within the philosophical world. According to him, there are philosophical "superpowers" – he mentions England, France and Germany – and there are philosophical "provinces", Poland belonging to the latter. Inevitably, he thought, superpowers are bound to draw provinces towards them. Hence, one crucial problem for him was to define the way in which "provincial" philosophy should relate to the ideas and systems coming from philosophical Leviathans. Twardowski argued that the former should neither seek to cloister themselves nor to remain subordinate to the latter. Provinces, on the contrary, have a privileged position in that they can make use of these various sources in their attempts at shaping their own original contributions to philosophy. Hence, he insisted that his students keep up with contemporary philosophical research and engage with "novelties" such as the mathematical logic of Frege and Russell or French conventionalism, for instance. On the other hand, he believed that philosophical training should provide students with a substantial insight into the history of philosophy as a whole. According to Twardowski, a rigorous scholarly analysis of foreign ideas supplemented by historical proficiency would support the production of original ideas. He recognised the latter to be, of course, neither sufficient nor necessary conditions for successful philosophical achievements and that results would ultimately depend on individual talents. Although Twardowski's approach might have encouraged a certain kind of pluralism, in favouring analytic reflection and logical rigor, he succeeded in shielding his students from eclecticism.

Twardowski's success was enormous. Thanks to his efforts, Polish philosophy soon became not only a significant and authoritative philosophical movement but also an international player.[6] At the heart of what would be come the Lvov-Warsaw School (LWS), we find mathematical logicians and philosophers of science but, as the following collection of essays shows, Polish philosophy in and around LWS extended to all philosophical fields.[7] Indeed, some of Twardowski's students whose most well-known contributions were in logic and the philosophy of science also explored less "orthodox" problems. Henryk Mehlberg, for instance, is best known for his contributions to the philosopy of science, but his *Time, Causality and The Quantum Theory* in fact contains many chapters devoted to the mind-body problem. As Urszula Żegleń aptly shows in her essay, Mehlberg worked with a sophisticated metaphysical framework which enabled him to develop, as soon as 1937, what might count as one of the most original approach to the problem multiple realization.

---

[6] See J. Skolimowski, *Polish Analytical Philosophy*, Routledge and Kegan Paul, London, 1967; J. Woleński, *Logic and Philosophy in the Lvov-Warsaw School*, Dordrecht, Kluwer Academic Publishers, 1989. The latter contains an extensive bibliography and we therefore restrict our selection here and in the remainder of the text to books published after 1989. For a general survey of the Lvov-Warsaw School, see *Polish Scientific Philosophy: The Lvov-Warsaw School*, F. Coniglione, R. Poli and J. Woleński (eds), Amsterdam, Rodopi,1993; *The Lvov-Warsaw School and Contemporary Philosophy*, K. Kijania-Placek and J Woleński (eds.), Dordrecht, Kluwer Academic Publishers, 1998.

[7] Twardowski himself was not a logician. He was interested in logic, repeatedly lectured on logical problems, but considered logic to be a mere instrument in the service of philosophy.

This being said, LWS's famed association with logic is of course far from being an exaggeration. For amongst the members of LWS many were indeed logicians, and indeed some of the most gifted and influential logicians of the last century. The first generation of Twardowski's students alone, those who graduated before 1914, includes Jan Łukasiewicz (1878–1956; in Warsaw after 1918), Stanisław Leśniewski (1886–1939; in Warsaw after 1918), Tadeusz Kotarbiński (1886–1981; in Warsaw after 1918), Zygmunt Zawirski (1882–1948; in Poznań after 1928, in Cracow after 1937), Tadeusz Czeżowski (1889–1981; in Vilna after 1918, in Toruń after 1945) and Kazimierz Ajdukiewicz (1890–1963; in Lvov, in Poznań and Warsaw after 1945). A considerable number of Twardowski's students obtained jobs in Warsaw and this explains why, soon after World War I, the School could grow a second branch. In Lvov, Twardowski and Ajdukiewicz went on to teach Izydora Dąmbska (1904–1983; after 1957 in Cracow), Maria Lutman-Kokoszyńska (1905–1978; in Wrocław after 1945) and Henryk Mehlberg (1904–1978; in Chicago and Toronto after 1945).

LWS's collaboration with the leaders of the Polish Mathematical School, notably Wacław Sierpiński and Stefan Mazurkiewicz was crucial. In his contribution to this volume, Roman Murawski discusses the mutual influence that characterised their relationships. He describes a climate of open exchange and convivial cooperation where mathematicians and philosophers understood and respected each other. Mathematicians in Poland were convinced of the benefits of philosophy – a rarity. Philosophers reaped rewards from this synergy as well. In particular, Łukasiewicz's and Leśniewski's collaborations with the chief players of the Mathematical School soon begot LWS's most illustrious progeny: the Warsaw Logical School. The latter would not only be the young Alfred Tarski's (1901–1983; in Berkeley after 1945) formative milieu, but served as the training ground for a series of other significant Polish logicians, a number of whom eventually found their ways in some of the most important intellectual centres in Anglo-American Academia: Henryk Hiż (1917–2007; in Philadelphia after 1945), Stanisław Jaśkowski (1906–1965; in Toruń after 1945), Czesław Lejewski (1913–2001; in Manchester after 1945), Adolf Lindenbaum (1904-c.1941), Andrzej Mostowski (1916–1975), Jerzy Słupecki (1904–1987; in Wrocław after 1945), Bolesław Sobociński (1906–1980; in Notre Dame University after 1945) and Mordechaj Wajsberg (1902-c.1943). Kotarbiński also gathered around him a number of students: Jan Drewnowski (1896–1978), Janina Hosiasson-Lindenbaum (1899–1942), Maria Ossowska (1896–1974) and her husband Stanisław Ossowski (1897–1963) who later contributed to ethics and the philosophy of social sciences, Edward Poznański (1901–1976; who moved to Jerusalem in 1938) and Dina Sztejnbarg, later Janina Kotarbińska (1901–1996). Jan Kalicki (1922–1953; in Berkeley after 1945) who graduated in the clandestine University of Warsaw during World War II was the last philosopher to have been formed within LWS.

LWS significantly influenced Polish catholic philosophy. In 1936, Drewnowski, Sobociński, Józef M. Bocheński (1901–1995; in Fribourg, Switzerland after 1945) and Jan Salamucha (1903–1944) formed a group that would come to be known as the Cracow Circle. Their aspiration consisted in applying analytic methods, including symbolic logic, to the modernization of catholic philosophy.[8] As Roger Pouivet argues in his article, the members of the Cracow Circle anticipated one of the most important aspects of contemporary metaphysics: the use of logical and linguistic analysis in the treatment of metaphysical questions. In doing so, they also opened the door to a new approach to theological problems and the relation between faith and knowledge. In his essay, which focuses on Salamucha's account of the relation between theology and philosophy, Pouivet offers an important contribution to the history of analytical thomism *avant la lettre*, that is, before Gilson, Maritain, Peter Geach and Elizabeth Anscombe, for instance.

The most spectacular achievements of LWS belong to mathematical logic and include, for one, Łukasiewicz's many-valued logics and the modal systems associated with them (to which Słupecki, Sobociński, Tarski and Wajsberg also contributed). Grzegorz Malinowski offers a broad reflection on the birth of many-valued logic in the work of Łukasiewicz, proposing an account of the latter's motivations for introducing a third semantic value and discussing the consequences and limits of this move. He reviews some recent interpretations of Łukasiewicz' system and extends his reflection to a more general assessment of the import of multi-valuedness as a logical trend in contemporary logic. A partial list of LWS numerous other logical innovations would include Leśniewski's systems of Protothetics, Ontology and Mereology (which reverberate in the writings of Sobociński, Tarski and Lejewski), Tarski's metamathematics (see also Lindenbaum and Kalicki), Jaśkowski's system of natural deduction and intuitionistic logic (see also Łukasiewicz, Tarski, Wajsberg), Tarski's formal semantics and the semantic definition of truth, Mostowski's theory of generalized quantifiers, Ajdukiewicz's categorial grammar, Hiż's application of logic to linguistics and a revolutionary approach to the history of logic. The latter was worked out more or less in concert by Łukasiewicz, Czeżowski, Bocheński and Salamucha and consists in assessing the development of logic in reference to the current state of the art. This approach led, among other things, to the rediscovery of Stoic logic and to some potent reconstructions of Aristotelian logic.

Polish efforts in the field of logic and formal semantics were indeed considerable. Five contributions are here supplementing an area of scholarship that has unjustifiably remained at the margins of recent research in the history of analytical

---

[8] At its peak in 1938, LWS counted more than 80 members. Information in brackets about affiliation is general and incomplete. See also : S. Leśniewski, *Collected Works*, Dordrecht, Kluwer Academic Publishers, 1992; *Leśniewski's Systems. Protothetic*, J. Srzednicki and S. Stachniak (eds.), Dordrecht, Kluwer Academic Publishers, 1998; *Tadeusz Kotarbiński. Logic, Semantics and Ontology*, J. Woleński (ed.), Dordrecht, Kluwer Academic Publishers, 1990; Z. Zawirski, *Selected Writings on Time, Logic and The Methodology of Science*, Dordrecht, Kluwer Academic Publishers, 1994; T. Czeżowski, *Knowledge, Science and Values. A Program of Scientific Philosophy*, Rodopi, Amsterdam, 2000; *The Heritage of Kazimierz Ajdukiewicz*, V. Sinisi and J. Woleński (eds.), Rodopi, Amsterdam, 1994; *Alfred Tarski and the Vienna Circle. Austro-Polish Connections in Logical Empiricism*, E. Köhler and J. Woleński (eds.), Dordrecht, Kluwer Academic Publishers, 1999; A. Burdman Feferman, S. Feferman, *Alfred Tarski Life and Logic*, Cambridge, Cambridge University Press, 2004; J. Salamucha, *Knowledge and Faith*, Amsterdam, Rodopi, 2003.

philosophy. Jan Woleński's contribution brushes a portray of the developments of semantics in Poland and offers an insight into the peculiarities of its practice over different periods. Emphasising the double thrust of Brentano and Bolzano – the latter's role in Polish thought is often underplayed – he takes us on a narrative that throws light on the problems and issues that eventually led to Tarski's unparalleled breakthrough. He also explains that it is Tarski's distinctively Polish semantical approach that made it possible for his theories to bear fruit where previous syntactic approaches could not.

While Tarski's inaugural work in metamathematics has been at the center of much debate, he has also come to be viewed as a logician whose interest in philosophical questions remained marginal and whose influence on the philosophers of his time – and on philosophers of language in particular – was in fact negligible.[9] In their contributions, Douglas Patterson and Paolo Mancosu both attempt to adjust this somewhat misguided impression. Patterson, for one, offers and in-depth analysis of Tarski's conception of meaning whose purpose is to vindicate Tarski's definition of truth against some recent attacks – by Putnam, among others – as well as the tradition to which he belongs. Mancosu, relying mostly on unpublished material, documents three important philosophical problems which engaged Tarski with the Logical Positivists: the metaphysical underpinnings of semantics, physicalism and the possibility of a nominalistic account of mathematics and science. To mention only one case, the debate between Tarski and Neurath on the metaphysical import of semantics is a valuable (and compelling!) contribution to the sociology of knowledge: because it did not create consensus among the Logical Positivists, the issue left them – most likely because of Neurath's adamant requests to that effect – with no option but to deliberately... avoid any public discussion of the topic.

Tarski, like many of his colleagues, gave part of the merit for their accomplishments to their teacher Stanisław Leśniewski and two contributions are devoted to the latter. Arianna Betti, who has discussed Leśniewski's impact on his exceptional student elsewhere[10], provides in her essay a novel insight into the source of Leśniewski's own peculiar approach to deductive systems. She argues that his theories were informed by his tacit, yet uncontroversially orthodox Aristotelianism. Denis Miéville, on his part, offers a concise presentation of Leśniewski's syntax and argues that, by contrast with logics that are developed on the basis of a defined vocabulary and recursive construction rules, Leśniewski's method generates ever more sophisticated logical functors thus offering a way to reflect better the logical complexity of cognitive processes.

Polish logical achievements informed a number of philosophical theories. Kotarbiński's reism, the ontological thesis according to which there exist only things – and the related semantic claim that all genuine names refer to *concreta* – is

---

[9] Here, of course, Donald Davidson is an important exception for his programme of a semantic theory for natural languages is explicitly based on Tarski's definition of truth.

[10] See Betti, Arianna, "On Tarski's Lesniewskian Background and the Ajdukiewicz Connection" in *New Essays on Tarski and Philosophy*, Douglas Patterson (ed.), Oxford, Oxford University Press, 2008.

one. Ajdukiewicz's radical conventionalism, according to which the affirmation and rejection of statements depend ultimately on meaning postulates is another. In her contribution, Anna Jedynak carefully and systematically distinguishes the different versions and implications of classical conventionalism as it is found in the works of Duhem, Poincaré and Le Roy in order to assess Ajdukiewicz's own radical version of conventionalism as well as its development. As Jedynak explains, Ajdukiewicz's radical conventionalism yields a framework for the history of science that seeks to account, among other things, for the rapid and often tidal nature of scientific revolutions. She explains that, according to Ajdukiewicz, the modification of one thesis or meaning-rule within a theory may precipitate the acceptance of other meaning changes and hasten the creation of a new theory. According to Ajdukiewicz, this was the case with the birth of non-Euclidean geometry and Einsteinian physics, which were both suddenly brought into motion by modifications of semantic conventions that were initially relatively minimal.

The philosophy of science and philosophical semantics were the cradle of a number of innovations: Hosiasson-Lindenbaum's logic of confirmation, Ajdukiewicz's theory of fallible modes of reasoning, Czeżowski's probabilistic theory of induction, Łukasiewicz's anti-inductionism, Zawirski's and Mehlberg's investigations in the philosophy of time and of quantum mechanics (independently), the Ossowskis philosophy of social sciences, Poznański's development of operationalism, Dąmbska's attempt at an elucidation of the concept of scientific law, Janina Kotarbińska's theory of ostensive definitions and Lutman-Kokoszyńska's various contributions to an elucidation of the concept of absolute truth (following ideas of Twardowski).[11] The Warsaw School of Logic was one of the most formidable and creative group in the entire history of the field. Whilst some of their ideas were overlooked, some did influence the international philosophical community. They, in any case, invariably compare with views developed in other intellectual centres and often bear with them great anticipations such as, for instance, physicalism (Kotarbiński's reism) and Popper's fallibilism (Łukasiewicz's anti-inductionism). In particular, LWS was, so to say, the Polish counterpart to the Vienna Circle and, indeed, one of its keenest interlocutor. The two schools were contemporaneous, they both put forward a strongly anti-metaphysical attitude and both very strongly believed that applying logic to philosophical problems would bear fruit. However, the Poles did not reject metaphysics. They had different views as regards the extent to which metaphysical and ontological problems could be rigorously dealt with.[12] They contended that at least *some* carefully stated metaphysical problems could be dealt with scientifically. This, indeed might explain the disagreement between Tarski and Neurath concerning the presumably metaphysical character of the notion of truth (see Mancosu's contribution, infra). As a whole, it would be no exaggeration to say that, in matters metaphysical, LWS was much more discriminating than Logical Empiricism.

---

[11] This is a very broad and equally selective survey.

[12] Since metaphysics did not have good press in analytic philosophy of the first half of 20th century, LWS actually preferred the label "ontology".

While the Vienna Circle maintained a rather uniform – and prohibitive – view on metaphysical questions, the members of LWS developed a variety of original (though sometimes incoherent) ontological positions. While Twardowski, for one, admitted intentional objects, Kotarbiński defended nominalism. Tarski had reistic inclinations, but his work in the foundations of mathematics led him to adopt a sort of perfunctory methodological Platonism. Łukasiewicz, by contrast, tended towards fully-fledged ontological Platonism, but Ajdukiewicz abstained from ontological commitments, arguing that metaphysics is only secondary to epistemology. Although all members of LWS accepted "genetic" empiricism, that is, the thesis that all knowledge is derived from experience, they differed in the detail of their positions. Leśniewski, Łukasiewicz and Tarski considered logic to be somehow rooted in experience, but Ajdukiewicz, before he later converted to radical empiricism, held a somewhat more Kantian view about the nature logic, which he derived from semantic considerations. Misgivings concerning the existence of sharp boundaries between analytic and synthetic sentences was a corollary of this brand of empiricism. In this, the Poles can be seen to have been led to various anticipations of Quine's view – though none of them would have followed him in committing to radical semantic holism. In ethical matters, Polish cognitivism was at odds with Viennese emotivism, at least in its radical form. Polish analytic philosophy was in general much less radical than early Logical Empiricism and closer to the more moderate stand the latter took after 1945.

The Lvov Warsaw School, whilst dominant in Poland between 1918 and 1939, did not exhaust the whole of Polish philosophy during this period. It is in the nature of any collection of essays to do injustice to an ambitious topic and this one is unfortunately no exception. Some important actors of LWS such as Władysław Witwicki and Władysław Tatarkiewicz will not be discussed in what follows, the main reason for this being that we chose to focus on epistemology and logic. Besides, we have deliberately omitted contributors to Polish philosophy who did not stem from Twardowski's school (and the list is by no means exhaustive): Leon Petrażycki, one of the greatest legal philosophers of 20th century; Bronisław Malinowski, an anthropologist who, it could be argued, founded functionalism; Florian Znaniecki, the sociologist who coined the notion of "humanistic coefficient" in social sciences; Henryk Elzenberg, an exceptionally original thinker particularly active in ethics and aesthetics; Ludwik Fleck, a forerunner of the historical school in the philosophy of science; Stanisław Ignacy Witkiewicz also known as Witkacy, a philosopher, writer and painter who was an important part of Polish intellectual life at the time. Finally, "neo-messianism", a doctrine developed by a group of philosophers of nature in Cracow, will also remain untouched.[13]

We allowed two noticeable exceptions. The first one is Leon Chwistek who studied mathematics and philosophy at the Jagiellonian University in Cracow – he was

[13] Cf. M. Gogacz, E. Nieznański, T. Ślipko, "Polen", in *Geschichte der philosohischen Traditionen Osteuropas*, H. Dahm and A. Ignatov (eds.), Darmstadt, Wissenschaftliche Buchgesselschaft, 1996, 257–386.

also trained in the Fine Arts – and later with Hilbert and Poincaré. As Bernard Linsky notes in his meticulous and well documented contribution to this volume, Chwistek work, mainly because it did not rest on an extensional approach to logic, was unlikely to capture the attention of his colleagues in Lvov and Warsaw. Yet, he was esteemed by Russell who examined some of his proposals in *Principia Mathematica* and whose support ultimately won – at Tarski's expense! – a chair in Lvov in 1929. Linsky offers a choice piece of scholarship towards a better understanding of the import of Polish mathematics for Russell's thought in the 1920s.

The second exception is Konstanty Michalski, a world renowned Medievalist who stemmed from the Cracow School where, in the 1910s, history established itself as the dominant academic subject. As Claude Panaccio explains in his examination of Michalski's contribution to Ockhamist studies, he was among the first to have exposed the history of philosophy in the Middle ages to rigorous scholarship and to have undertook the painstaking work which it involves. Michalski devoted himself to identifying, classifying, attributing, dating and even summarizing a remarkable number of manuscripts. In this, he helped shape the methodology of modern Medieval Studies. In his paper, Panaccio examines some of Michalski's claims about Ockham, thus illustrating the unabating relevance of Michalski's work for the field as a whole.

Though Roman Ingarden can hardly be counted among the members of LWS, his case is at any rate not as markedly exceptional as the preceding two. Ingarden (1890–1970; after 1945 in Cracow) did begin his studies with Twardowski. But he moved to Göttingen after only one semester: Twardowski had sent the promising pupil there to study with the famous psychologist G.E. Müller. But Ingarden soon developed a fascination for Müller's colleague, Edmund Husserl, and the latter's "new" philosophical approach – phenomenology – and became his student. This, at any rate, clearly situates him, just as most other students of Twardowski, in Brentano's direct lineage. As Dariusz Łukasiewicz argues, what distinguishes him from his Polish colleagues, however, is the fact that while he came to harshly criticize Husserl's later transcendental subjectivism, throughout his career, Ingarden nonetheless remained faithful to the principles of Husserl's early realist phenomenology.[14] In 1933 Ingarden replaced Twardowski in Lvov. He won only few students over to phenomenology, but those who were, seemingly also adopted

[14] Unfortunately, there is no detailed survey of Ingarden's philosophy. One may consult *Roman Ingarden and Contemporary Polish Aestethics*, P. Graff and S. Krzemień-Ojak (eds), Warsaw, Polish Scientific Publishers, 1975; G. Haefliger, *Über Existenz: Die Ontologie Roman Ingardens*, Dordrecht, Kluwer Academic Publishers, 1994; *Roman Ingarden i filozofia naszego czasu* (Roman Ingarden and the Philosophy of Our Times), A. Węgrzecki (ed), Cracow, Polskie Towarzystwo Filozoficzne, 1995 (the latter contains several papers in English and German); A. Chrudzimski, *Die Erkenntnistheorie von Roman Ingarden*, Kluwer Academic Publishers 1999. See also the encyclopedic articles on Ingarden (by A. Przyłębski) and Poland (by K. Górniak-Kocikowska) in the *Encyclopedia of Philosophy*, L. Embree, E. A. Benhke, D. Carr, J. C. Evans, J. Huertas-Jourda, J. J. Kockelmans, W. R. McKenna, A. Mickunas, J. N. Mohanty, T. S. Seebohm and R. M. Zaner (eds.), Dordrecht, Kluwer Academic Publishers, 1997.

Ingarden's deeply rooted realism. Ingarden's ambition was to create a centre for phenomenological research in Lvov and he might have succeeded if the war had not brought his efforts to an end. Ingarden should be regarded as having attempted for the study of mind in Poland what Twardowski had done for the study of truth and knowledge.[15]

Leopold Blaustein (1905–1944) who studied for a time with Ingarden, was also a Polish phenomenologist. By contrast with Ingarden, and although Blaustein also studied with Husserl in Freiburg im Breisgau in 1925, Blaustein's thought was substantially shaped by Twardowski who was supervising his doctoral work. Blaustein's approach to questions about the mind and consciousness nonetheless stands out. His work on perception, for instance stands in close proximity with the cognitivsm of the *Gestalt* psychologists with whom he came in contact at Köhler's institute during an academic stay in Berlin. In her essay on Blaustein, Wioletta Miśkiewicz argues that by enriching the Husserlian notion of mental content, Blaustein comes close to a theory of perception which is both more accurate than Husserl's and also more amenable to contemporary cognitive theories. Phenomenology must be counted as the second most important movement in 20th century Polish philosophy.

World War II was as dramatic for Polish philosophy as it was for Poland in general. Many intellectuals died, most often at the hands of the Nazis, especially Jews. Despite the fairly reliable operations of clandestine universities, education and scientific activities were conducted under substandard conditions. Many philosophers left Poland. The reconstruction of Polish philosophy after 1945 was however surprisingly rapid. Poland soon resumed its pre-war activities, that is, momentarily, until 1948 when the newly established communist government began to introduce a series of constraints, the covert purpose of which was to impose Marxist intellectual hegemony. Interestingly, despite considerable political and ideological pressure, Polish philosophical life did not become mere pretense. Polish philosophy did preserve its pluralistic character in the most difficult of times for, ironically, despite the efforts of administrative authorities, Marxism rather than becoming an instrument of conversion, merely became a form of academic philosophy among others. This phenomenon, unique in Eastern Europe, contributed to a large extent to the anti-Communist revolts of 1989.

Today, more than two decades after the collapse of most communist regimes in Eastern Europe, Polish philosophy is finally being recognised for what it is: a locus of scientific innovation whose only barrier might have been linguistic. For the historian of analytical philosophy, there are no longer any excuse and the recognition of the importance of LWS's place in the intellectual landscape of the twentieth century comes with a responsibility: avoiding the mistake which would consists in denying it one's scrutiny on the pretext that it had little influence on the course of

[15] The relations between LWS and Ingarden were somehow tensed. He was skeptical about the importance of mathematical logic for philosophy. On the other hand, logicians disliked his wild metaphysics. Yet his presence in Poland had the beneficial effect of inciting Polish philosophers to better understand traditional philosophical problems, such as, for instance, the debate between idealism and realism.

mainstream philosophy. This would be a double mistake: it would be plainly false since some of the greatest achievements of the last century, the development of many-valued logics, mereology, categorial grammars and the semantic definitions of logical consequence and truth, to name only those who have an abiding impact on the way philosophy is done today, have their origins in the thought of some of LWS's most prominent thinkers. It would be a mistake, on the other hand, because the heuristic potential of any given episode in our philosophical history cannot be assessed *a priori*.

We have to know our past in order to know ourselves and the history of philosophy in Poland is also our history, whether we are Polish or not.

Part I
Twardowski and Polish Scientific Philosophy

Chapter 1
Polish Metaphysics and the Brentanian Tradition

Dariusz Łukasiewicz

1.1 Introduction

There is a view according to which the philosophy of the Lvov-Warsaw School is rooted in the tradition of British analytic philosophy and is closely linked to logical empiricism. Although this is not entirely false – both current had repercussions in the work of Polish philosophers – I wish to argue that the best way to understand and evaluate the most significant achievements of philosophy in Poland is to consider it in close relation to the Brentanian tradition, on the background of Brentano's and Twardowski's views, in particular. Brentano's impact was twofold. It consisted, firstly, in the fact that his "scientific" style of philosophizing turned out to be very attractive for Twardowski and his followers in Poland, and, secondly, in the fact that he rejected Kant's heritage and returned to the tenets of classical philosophy. But it would be hard to imagine the development of Polish philosophy without Twardowski's essential modifications of Brentano's main ideas. Kazimierz Ajdukiewicz in 1937 in a discussion of the particular influence of Brentano's ideas on the philosophy of the time (Ajdukiewicz 1985a, 252), distinguished four essential elements in Brentano's philosophy: (i) the theory of intentionality, (ii) the "pre-Kantian" theory of truth (i.e. the classical, Aristotelian conception of truth), (iii) ontological realism and (iv) the commitment to a theistic metaphysics. I will document the fact that all four Brentanian elements, though natural theology to a lesser extent, were embraced by at least some Polish philosophers – I focus on Stanisław Leśniewski, Tadeusz Kotarbiński, Tadeusz Czeżowski, Kazimierz Ajdukiewicz and Roman Ingarden – and that these ideas provided the foundation for some of their most significant and original achievements.

1.2 Twardowski on Intentionality and Truth

Twardowski rejected Brentano's immanentism. Intentional relations obtain, Twardowski argued, between thoughts, that is, mental acts and external objects and

D. Łukasiewicz (✉)
Kazimierz Wielki University, Bydgoszcz, Poland
e-mail: darek.lukasiewicz@interia.pl

S. Lapointe et al. (eds.), *The Golden Age of Polish Philosophy*, Logic, Epistemology, and the Unity of Science 16, DOI 10.1007/978-90-481-2401-5_2,

not, as Brentano maintained between thoughts and (mentally) immanent objects. Twardowski's amendment to Brentanian immanent intentionality had at least two fundamental consequences. It allowed Twardowski to retain the study of intentional objects as the basis for formal ontology and thus to understand the latter as a "*daseinsfreie Wissenschaft*": the study of intentional objects does not imply any commitment as regards real objects. For Twardowski, immanent entities are not objects but "contents" sharply distinct from the acts of presentation and their referent. Twardowski, following Brentano, conceived objects as the possible correlate of a presentation, but contrary to Brentano, claimed that an object's being an object of presentation does not entail its existing in any real or possible way (Twardowski 1982, 27). It also allowed him to set up semantics as the theory of the relations between thoughts and language, on the one hand, and external reality on the other. According to Twardowski, since the mind, through its contents, is intentionally related to external objects, and since the main role of language is to intimate (*kundgeben*) what is in the mind, then linguistic signs are also in some substantial way related to external objects. In distinguishing mental actions from their products (on this, see Bobryk this volume), Twardowski is furthermore in a position to grant stability to intentional and semantic relations: the contents involved in intentional and semantic relations are to be conceived as the stable products of (certain aspect of events that take place in) the mind and not individual, subjective, perishing entities. As a result, Twardowski could easily accept that truth is a relation between thought and external reality and approve of ontological realism.

Twardowski did for a time vindicate the existential theory of judgment proposed by Brentano. Unlike Brentano, however, Twardowski also forcefully defended the idea that truth is absolute. Brentano believed that while a judgment may on the face of it appear to remain the same, its content may change from occasion to occasion, or from subject to subject (Brentano 1930, 26). This was a consequence of Brentano's immanentism, according to which acts of judgments are conceived as real events in a way that leaves no room for any view of truth and falsity as timeless properties. One can find important consequences of Twardowski's absolutism both in Leśniewski's and Kotarbiński's ontological views, as we will see, as well as in Tarski's semantic definition of truth.

1.3 Leśniewski's Ontology

"Ontology" is the name for a logical calculus which Leśniewski interpreted as an extended modern version of the traditional logic of names. Leśniewski's Ontology is rooted in Twardowski's theory of objects, which finds its roots in Brentano's. Leśniewski's aim is to establish nominalism and to express it in a precise logical language. Ontology is to be such a language. Leśniewski's nominalism can be regarded as a result of his criticism of Twardowski's doctrine of "general" objects. It is not clear why Leśniewski started with the analysis of general objects: as Woleński suggests, it might have been the result of Anton Marty's inspiration, whose major work Leśniewski knew and even intended to translate from German into

Polish (Woleński 1990, 215). At any rate, according to Leśniewski – but Tadeusz Kotarbiński, for instance, also subscribed to this view – Ontology is to be a theory of the general principles of being (Kotarbiński 1966, 210). Ontology was meant to meet some of the desiderata Twardowski had expressed in his paper "Symbolomania and Pragmatofobia" (Twardowski 1921) to the effect that science ought to deal not with symbols of things but with the things themselves. This accounts for Leśniewski's so called "intuitive formalism". Intuitive formalism is the view that the theorems of Ontology have determinate meanings and refer to objects: Ontology is to be a true description of the world. It is in this sense that it is to provide the general principles of being. In this respect it is not unlike Russell's early conception of logic as the formal theory of the real world (Cf. Smith 1990, 140).

Ontology is based on a unique axiom[1]:

$$\forall AB\ [A \text{ est } B \Leftrightarrow \forall x\ (x \text{ est } A \Rightarrow x \text{ est } B) \wedge \exists x\ (x \text{ est } A) \wedge \forall xy\ (x \text{ est } A) \wedge y \text{ est } A \Rightarrow x \text{ est } y)].$$

Kotarbiński explained the sense of this axiom as follows (Kotarbiński 1966, 190–191):

> For any *A* and *B*, *A* is *B*, if and only if: (i) for any *x* and *y*, if *x* is *A*, then *x* is *B*; (ii) for some *x*, *x* is *A*; (iii) for any *x* and *y*, if *x* is *A* and *y* is *A*, then *x* is *y*.

Whatever terms are chosen to stand in the place of *A* and *B*, the sentence "*A* is *B*" is equivalent to the following conjunction:

(1) For all *x*, it is true that if its designatum falls under *A*; it falls under *B*;
(2) It is possible to choose a term for *x*, such that its designatum falls under *A*;
(3) For all *x* and *y*, it is true that if the designatum of *x* falls under *A* and the designatum of *y* falls under *A*, then the designatum of *x* is the designatum of *y*.

Leśniewski's axiom can be formulated in shorter form with the aid of the following definitions:

(a) $\forall A[\text{ex}A \Leftrightarrow \forall x(x \text{ est } A)]$ – the definition of "*A* exists".
(b) $\forall A[\text{ob}A \Leftrightarrow \exists x(A \text{ est } x)]$ – the definition of "*A* is an object".
(c) $\forall A[\text{solA} \Leftrightarrow \forall xy\ (x \text{ est } A(x \text{ est } A \wedge y \text{ est } A \Rightarrow x \text{ est } y)]$ – the definition of "There is at most one...".
By virtue of definitions (a)–(c), one can formulate the axiom of Leśniewski's Ontology as follows:
(d) $\forall A,\ B[A \text{ est } B \Leftrightarrow \forall x(x \text{ est } A \Rightarrow x \text{ est } B) \wedge \text{sol } A]$.

where "*A*" and "*B*" stand for any expression belonging to the category *name*. They may be as Leśniewski claims: (i) ordinary singular denotative names or nominal

[1] My presentation of Leśniewski's Ontology is very limited and based on Kotarbiński 1966, Küng 1963, Smith 1990 and Woleński 1999a. See also Betti's article in the present volume.

expressions, like "Socrates" or "Pope John Paul II"; (ii) common or general names, like "philosophers" or "cats"; (iii) fictional or empty singular names, like "Pegasus" or "the largest prime number"; or (iv) fictional or empty general names, like "sirens" or "round square". All such expressions belong to a single category because whether a nominal expression like "the woman at the door" is singular, or common, or empty, depends on the state of the world and is therefore not relevant from the logical standpoint.

(d) serves to translate sentences of natural language into theorems of Leśniewski's Ontology. For a singular sentence like "Socrates is a philosopher" the translation will take the form of the conjunction of three following sentences: (A) Socrates exists; (B) There is at most one Socrates; (C) Whatever is Socrates is a philosopher.

One thing of great significance to Leśniewski is the fact the functor "*est*" contained in the axiom above and in the theorems it implies do not have a temporal meaning; "*est*" is a temporal: it is meant to express the "absoluteness" of truth. Twardowski defended the view that truth is absolute and, later on, Leśniewski in his paper "Czy prawda jest tylko wieczna czy też i odwieczna?" (*Is truth only eternal or both eternal and sempiternal*) (1913) held the same position in his controversy with Kotarbiński's view presented in "Zagadnienie istnienia przyszłości" (*The problem of the existence of the future*) (1913). Moreover, the functor "*est*" has no existential meaning: this means that Leśniewski's Ontology is a "free logic".[2] Let us assume that the antecedent part of Leśniewski's axiom is true so that *A* exists and there is at most one *A*. From this, it follows that *A* is an object. Now, while Leśniewskian Ontology is not completely ontologically neutral – one of its theorem states that ob*A* exA sol*A* implies that if something is an object, then it is an individual – it is at any rate consistent with both reism according to which there are only things, as well as with set-theoretical Platonism, for example, according to which there are only sets (Cf. Küng 1963, 85).

1.4 Kotarbiński's Reism

Tadeusz Kotarbiński was reluctant to build philosophical systems. He nonetheless invented an ontological language which he considered to be a model of the real world, "reism", and he was convinced (not quite correctly) that the logical foundations of the latter were provided by Leśniewski's Ontology. Kotarbiński developed his reistic ontology over many years, and he himself identified many stages – seven in total! – in this development. Roughly, in the early years of his philosophical career he argued for a form of nominalism he called "concretism". Nominalism was a reaction to the rich pluralistic ontology defended by Twardowski and other Brentanians (like Meinong or Husserl). It may also have been partly inspired by Leśniewski's

[2] Woleński understands this to result from the following considerations. If we assume that there is no object, then both sides of Leśniewski's axiom are false, and, hence, the axiom by virtue of the meaning of the equivalence connector is true. Hence, the theorems of Ontology are true in the empty domain (Woleński 1999a, 21). So Leśniewskian Ontology is a '*daseinsfreie Wissenschaft*', like the Twardowskian theory of objects or Meinong's *Gegendstandstheorie*.

rejection of general objects (according to Leśniewski, the so-called general objects are contradictory and hence impossible). From nominalism, which says that every entity is an individual, neither reism (every entity is a thing) nor pansomatism (every thing is a body – the view Kotarbiński later came to adopt) follow. Kotarbiński came from nominalism to reism by rejecting the categories of what we may want to call classical Brentanian ontology, among them: immanent contents, properties, events, relations and states of affairs (Kotarbiński 1966, 326). As regard the latter comment one should note that, interestingly, while the actual starting point of Kotarbiński's criticism was the classifications proposed by Wilhelm Wundt (Wundt 1906, 113). Wundt had himself borrowed the latter from Brentano. Kotarbiński believed that the categories he rejected were neither indispensable nor did they fit common sense intuitions. He thought the concept of thing, by contrast, to be fully coherent with Aristotle's concept of first substance (Kotarbiński 1966, 326). Reism, in his radical somatistic form, claims that every entity is a thing and every thing is a body, that is, roughly speaking, spatio-temporal entities.[3]

As is well known, Franz Brentano abandoned his earlier ontological views around 1904 and came to believe that only things exist. We know that Kotarbiński was not directly acquainted with Brentano's theories and learned about Brentano's ideas through Twardowski at the end of the 1920s. But then, Kotarbiński became convinced that Brentano (as well as Leibniz) held views similar to his. According to Kotarbiński, the only crucial difference between Brentano and him consisted in the fact that Brentano accepted two kinds of things, bodies and souls, while Kotarbiński accepted only one: bodies. The soul, for Kotarbiński, is a sort of body. To be precise, there is more than one important difference between the two philosophers as regard the concept of thing. Brentano's concept of thing is rooted in his version of descriptive psychology. It is a concept applied without restrictions to objects of any ontological domain because, according to Brentano, objects of all categories may be objects of presentations. Brentano counts as objects the zero-dimensional entities he calls "souls" as well as the three-dimensional entities he calls "bodies". It is characteristic of Brentano, however, that material things are not restricted to three-dimensional physical bodies. They include topoïds of higher number of dimensions and embrace also things of lower number of dimensions such as boundaries, points, lines and surfaces. Kotarbiński, by contrast, resorts to a rather intuitive concept of thing which is based on examples taken from everyday life experience of bodies and described with the aid of such words as "bulky", "extended", "inert" (Kotarbiński 1966, 330). While Kotarbiński's ontology was neither directly inspired by Brentano nor indeed consistent with it, they both agreed that whatever exists is individual, and that ontological arguments should be based on semantic analyses.

In his 1935 paper "On Pansomatism", Kotarbiński argued that a thing is not a three-dimensional entity but a four-dimensional object: it is a whole which starts to exist, and then ceases to be at a given moment. According to this "mereological"

[3] The concept of a thing in Kotarbiński's reism was not univocal. In *Elementy* he understood by 'thing' a lasting, bulky and inert body. But in "On Pansomatism" Kotarbiński introduced the idea that a thing is a logical construct consisting of many distinct temporal and spatial parts.

concept of thing – as Tadeusz Czeżowski would later describe it – no individual is ever a whole at any particular moment (no individual is ever a unitary object). An object can only be regarded as such by considering it from the very beginning of its existence to the very end. Even the self, at a particular moment of its existence, is only a part – a piece or "slice" – of an individual whole (Czeżowski 1958, 167). Each thing can be divided into its temporal parts, somewhat like it can be divided into its spatial parts. Kotarbiński never developed his views on temporal parts; he merely introduced the notion in the above mentioned paper. It must nonetheless have been a crucial view for him: Kotarbiński presented it in private to Tadeusz Czeżowski, Czeżowski discussed the concept of temporal parts in his 1958 paper relying on Kotarbiński's explanations (Czeżowski 1958, 169). Kotarbiński himself presumably took over this view from Leśniewski who introduced the concept of an object's "time slices" in his "On the Foundations of Mathematics" of 1928. Leśniewski, on his part, came up with the idea of temporal parts as an answer to some of the paradoxical consequences resulting from the axiom of Ontology (Leśniewski 1992, 380–381). As we have seen above, the "*est*" we find in the axiom of Ontology has no temporal meaning. The timelessness of "*est*" results from Leśniewski's insistence on the absoluteness of truth attached to the theorems of Ontology. They are to be true independently of the time and context of their utterance. This was, let us emphasise it, almost certainly inspired by Twardowski who, under the direct influence of Bolzano's *Theory of Science*, had written a famous paper on the topic in 1894. In the sentences "Socrates is a philosopher" and "Socrates is seating" the word "is" has the same atemporal meaning as in the sentence "The number 3 is an odd number". The point is that the sentence "Socrates is seating" is about Socrates who is doing something at a particular time, namely now. But according to Leśniewski's Ontology, the reference to time occurs not through the functor "*est*" but through the subject of the sentence, i.e. Socrates. This can be made explicit if "Socrates" is equipped with a temporal index, for example, an advervial expression of the form "at t". The result of such an operation is a nominal expression of the form: "Socrates at t'."[4] Now, according to Leśniewski – Bolzano does not draw this consequence – the expression "Socrates at t stands for a certain temporal part of Socrates taken as a whole.

Leśniewski's and Kotarbiński's ontologies of temporal parts are consistent neither with Aristotle's conception of substance, nor with Brentano's concept of thing, which manifestly stayed closer to Aristotelian metaphysics. It was an original idea.[5] According to Aristotle, the whole thing exists at each particular moment of time. The enduring thing is not a logical construct and it is a unitary whole. There are, concerning the ontology of time, big differences between Brentano and Kotarbiński. On Kotarbiński's view, what passed exists as a past thing, and what will be, exists as a future thing. The realm of existing entities cannot be restricted to the presently

---

[4] As Sandra Lapointe pointed out this view can be traced back to Bolzano who was the first to propose it (Bolzano, 1837, §79, vol. 1, 365).

[5] Later on a similar view was defended by Woodger (Woodger 1939).

existing things. However, according to the later Brentano, all that exists is identical with what exists just now at the present moment, and Brentano's world therefore differs significantly from Kotarbiński's world. The latter it is a world in which there exists only one – albeit continuously changing – instant (cf. Smith 1990, 177). There are however also significant similarities between other aspects of Kotarbiński's and Brentano's ontologies. Firstly, both were convinced that two separate things may form a new thing: bodies of air, swarms of bees, the solar system and complex entities like societies, nations and social classes. The latter are all new entities formed by already existing things. Kotarbiński's view that collections (set, sums, the term need not be taken in a determinate technical sense for our present purpose) of things can themselves be counted as things is grounded in Leśniewski's mereology. On Leśniewski's account, if *a* and *b* are things, then their sum forms a new thing *c*, irrespective of whether *a* and *b* are connected, or contiguous, or materially related in any way; even irrespective of whether *a* and *b* exist at the same time. In this, Kotarbiński, like Brentano – but the former presumably under Leśniewski's direct influence – broke with Aristotle's ontology of substances. According to Aristotle, a substance has the unity of a living thing and, in consequence, a part of a thing, for as long as it remains a part, is not itself a thing, but only possibly so. It can be said that one can find, in Brentano (Brentano 1933, 53) as well as in Kotarbiński (Kotarbiński 1966, 434), a reformulation of Aristotle's doctrine of substance and accident in terms of part-whole relation. In many places of his *Elementy*, Kotarbiński provided examples which support such a conception. According to this theory, the word "is" in the sentence "Socrates is seating" has a different meaning than the "*est*" in the axiom of Leśniewski's Ontology. The word "is" expresses a part-whole relation. The word "seating" – or the complex expression "the seating Socrates" – is the name of a transitory existing thing in which Socrates himself is included as a part. On this reading, Socrates is a part of the seating Socrates. In Aristotle's terminology, this would amount to saying that a substance is always part of an accident. The advantage of this conception is that it accounts for the changes that take place in the world without having to resort to an ontology of temporal parts. The problem, however, is that on Brentano's and Kotarbiński's view, "the jumping Socrates" refers to a new thing although no second thing is added to Socrates to make him a jumping Socrates. So while this conception may be seen to avoid the consequence of objects' having temporal parts, it leads to the multiplication of things in the world: there are thinking Socrates, jumping Socrates and seating Socrates and they are all distinct things although they contain as their part the same substance, namely Socrates himself.

1.5 Ajdukiewicz's Ontology

Ajdukiewicz saw a deep analogy between semantics and epistemology: just like semantics shows how one can pass from statements about language *L* to statements concerning the things about which *L* speaks, Ajdukiewicz speculated that one can pass from statements about cognition to statements about the objects of cognition.

Following Tarski's specification for passing from sentences about L to sentences of L, Ajdukiewicz assumed that the transition to statements about objects was possible, provided that the language of epistemology was sufficiently rich and comprised both names for thoughts and names for things. (Ajdukiewicz 1948, 337). Although Ajdukiewicz didn't in the end draw ontological conclusions from his epistemological investigations, his results have obvious implications (cf. Przełęcki 1989, 55–56). For instance, if epistemological idealism is false (as follows from his assumption), then epistemological realism must be true, and this in turns means that the object of cognition is independent of cognition. Ajdukiewicz was convinced that epistemological realism is a good foundation for ontological realism and he wrote, in the context of the controversy between metaphysical idealism and realism, that his epistemological works led him to refuse all form of idealism and prompted him to accept a realistic position (Ajdukiewicz 2003, 93).

Ajdukiewicz's refutation of idealism is also rooted in the idea of an intentional yet non immmanentistic relation between thoughts and objects. In his argument, Ajdukiewicz resorts indirectly to the form of intentionality worked out by Twardowski – as opposed to Brentano and to the later Husserl. He writes:

> Philosophers often programmatically cut themselves off from the object-language using various devices such as 'epoche', 'Einklammerung', etc. The use of those devices amounts simply to abandoning the object-language for the duration of epistemological analysis and to restricting one's language to the language of syntax. The thesis argued in the present paper is that a philosopher who has so abandoned the object-language, i.e., the language which we normally use in everyday life to describe reality, will be unable to say anything about reality
>
> (Ajdukiewicz 1948, 347)

Ajdukiewicz strongly believed that epistemology is an indispensable foundation for metaphysics. He accepted not only the intentionality of mind but, like Brentano, such entities as acts, concepts, propositions, states of affairs, and values. Values, for him, are the correlates of (true or false) logical propositions that are based on adequate emotions (Ajdukiewicz 1985b, 346). The connection to Twardowski and Brentano, and Ajdukiewicz's general sympathy for their metaphysics is important: Ajdukiewicz is often regarded as a representative of logical empiricism rather than a representative of the Brentanian tradition in Poland. But this is an exaggeration.

1.6 Metaphysics and Philosophy of Religion in the Lvov-Warsaw School

Polish philosophers neither consistently nor systematically sought to carry out the Brentanian project according to which philosophy should also deal with the problems of natural theology. But none of the representatives of the Lvov-Warsaw School ever claimed that propositions dealing with the objects of the latter were meaningless either. Twardowski himself in the early stage of his philosophical career dealt with the problem of the immortality of the soul. He believed that in order to prove that the soul is immortal, it is both necessary and sufficient to analyse the data of inner experience (Twardowski 1895, 472). According to him, these data support the thesis of the unity and identity of the soul, which is understood as the (immaterial) subject of mental acts. The latter thesis can, in turn, justify the belief in the existence of a perpetual soul, and Twardowski reasoned that if it is perpetual, then the soul is also immortal. By the time he was giving his lectures on medieval philosophy in 1910, Twardowski however appears to have changed his mind about the immortality of the soul and about philosophy of religion in general (Twardowski 1910, 107). Traditional problems of the philosophy of religion, such as the existence of God, God's relation to the world, and the immortality of soul, he came to believe, could not be regarded as objects of rational knowledge but only as the subject matter of religious beliefs. In his opinion only formal ontology – that is "descriptive", as opposed to "speculative" metaphysics – has a truly scientific character. In the early 1930s he revised his views once more claiming that there is no gap between religious and scientific beliefs, and that the two spheres may even cooperate since religious systems sometimes comprise prescientific elements which science ought not to condemn or neglect (Twardowski 1965, 384).

Tadeusz Czeżowski agreed with Twardowski's postulate that science and religion ought to cooperate. Czeżowski claimed that statements that express religious beliefs, when they have been properly formulated, can be incorporated by scientific theories as antecedents of material implications whose consequences are laws (of nature) or other general statements.[6] Czeżowski also had views on the concept of God. According to him, the concept of God used in traditional speculative metaphysics is logically inconsistent because it entails contradictory properties, namely omnipotence and mercy. He criticised arguments for the existence of God as well, but conceded that while there are no correct proofs of God's existence, it is not excluded that some arguments could be worked out that are at least not logically inconsistent (Czeżowski 1948, 91). The Brentanian conception of religious philosophy found its most systematic realisation in the Cracow Circle (Bocheński, Drewnowski, Salamucha, and Sobociński) who were mainly inspired by the project of a scientific religious metaphysics put forward by Jan Łukasiewicz, according to whom the latter should use the axiomatic method.

For Kazimierz Ajdukiewicz while rationality is a basic criterion of scientific research, it should also be understood to have social significance: it should be used, in everyday activities, to preclude nonsense and false beliefs. In spite of such declarations, Ajdukiewicz was not an enemy of metaphysics in general. On the contrary, he believed that metaphysics regarded as a kind of world-view was indispensable as the foundation of ethics. In his view, moral evaluations are based on metaphysical beliefs of the sort we find in religion (Ajdukiewicz 2003, 145–147). Nonetheless, he thought it imperative that metaphysics have a scientific character in order to guarantee its own stability and, in consequence, the stability of moral values. This scientific character of metaphysics could be preserved only if the latter was based on rational epistemology.

[6] I argue for this in Łukasiewicz 2002, 212–220. See also Woleński 1989, 68–74.

1.7 Ingarden and Brentano

Roman Ingarden is usually regarded as a phenomenologist. He was one of Edmund Husserl's most distinguished student and remained, throughout his career, under the influence of Adolph Reinach and Alexander Pfänder – two of Husserl's other students. But Ingarden belongs within the Brentano-Twardowski filiation as well. He started as Twardowski's student and it is Twardowski who recommended that he go to Halle where, instead of studying under the psychologist Müller as Twardowski had hoped, he fell under Husserl's influence. He studied Brentano's works and, in 1936, gave lectures on Brentano's philosophy at the University of Lvov. It was Ingarden who held the speech at Twardowski's funeral in 1938. Ingarden was acquainted with the detail of Twardowski's theory of general objects (Ingarden 1966, 21–35); he knew and often criticised the theories of Twardowski's disciples and regarded the latter as representatives of Logical Empiricism in Poland. He also had passionate discussions with Leśniewski and Kotarbiński on the question of general objects, and with Czeżowski on the ontological argument he put forward in *The Controversy over the Existence of the World* (Ingarden 1951). In addition to those external historical connections there is at least one other interesting fact about Ingarden: all essential constituents of the Brentanian philosophical project are realised by Ingarden and to such an extent that the idea that his philosophy is closer to Brentano's than Husserl's phenomenology, for instance, should not be regarded as implausible. This of course, does not imply that Ingarden was a fully-fledged "member" of the Lvov-Warsaw School – the latter claim could not be sustained. But Ingarden was certainly at home amongst the Polish philosophers of his time.

Ingarden, like Husserl and Twardowski, rejected Brentano's early immanentism and distinguished two kinds of intentional objects: the intentional object as a target of intention and the intentional object as a product of intention (while they can sometimes be the same, they need not be).[7] Ingarden's conception of intentional objects as products of mental activity is very similar to Twardowski's (see also Bobryk, this volume). Ingarden's theory of intentional objects is nonetheless highly original: for one thing, as products, intentional objects are taken to have a special mode of existence, namely intentional existence. In this respect, as far as the ontology of meaning is concerned, Ingarden, contrary to Husserl, followed Twardowski (and Brentano), who rejected Bolzanian-type Platonism. However, Ingarden, like Husserl, Reinach and Meinong developed a sophisticated ontology of truth based on the category of states of affairs. To this effect, he firmly rejected the existential theory of judgement Brentano had put forward and which had been defended by Twardowski and Czeżowski among others.[8] Besides, like Brentano and Twardowski, he disputed subjective idealism and the latter was the main object of his well-known controversy with Husserl.

---

[7] For more on this question, see. Chrudzimski 1999, 81–92.

[8] Compare D. Łukasiewicz 2006 and also D. Łukasiewicz 2007.

Ingarden worked out a rich system of ontological categories embracing among other things individual objects, general objects (ideas), states of affairs, events, processes, properties, relations and values. His defence of realism about the external world was original in the context of the Brentanian tradition: he analysed intentional objects, i.e. the products of the human mind, and saw no problem in comparing them with the real objects of common sense experience. Ingarden is well known for his analyses of literary works, but he was interested in other kinds of human "products". The path from the theory of art to ontology is an unusual one. In *The Controversy over the Existence of the World*, which is concerned with the idealism–realism debate, Ingarden analysed concepts, or, as he preferred to, say "ideas" of various kinds of objects (including the idea of existence) (Ingarden 1964/65, 77). The conclusion of Ingarden's ontological investigations is that neither idealism (whatever the variety), nor the kind of absolute realism proclaiming the ontological primacy of the real world, are possible. His approach was criticised by Tadeusz Czeżowski, who argued that mere conceptual analysis cannot result in propositions asserting the existence (or inexistence) of the objects falling under any of these concepts (Czeżowski 1958, 37).

One must also remember that Ingarden was one of the first in Poland to pay attention to the fact that Brentano was not only a psychologist, but had also tackled epistemology and philosophy of religion (Ingarden 1969, 465). In fact, just like Brentano, Ingarden subscribed to a form of rational theistic metaphysics. The conclusions he drew from ontological considerations were indeed pretty clear: if there is a world, it is independent of the human mind, but it is nonetheless created by a "pure consciousness". He thought it highly unlikely that such a pure consciousness could be obtained through some kind of "phenomenological reduction" à la Husserl. He believed that one should assume some other kind of pure consciousness and some other kind of (non human) subject linked with it (Ingarden 1964/65, 585). Hence, it seems that the core of the controversy between idealism and realism – the question of the status of consciousness and its relation to the world – is bound with a certain conception of God, and with the metaphysical theology to which certain kinds of ontological studies sometimes lead.

Ingarden knew Brentano's philosophy and his coming close to Brentano's project of a theistic metaphysics may not have been a mere coincidence. Ingarden's theistic-metaphysical spirit is however perhaps best visible in his aesthetic writings, where he discusses metaphysical qualities such as sinfulness, holiness and meanness. He believed that such properties cannot be rationally grasped but that through them one can reveal the deeper meaning and source of being (Ingarden 1973, 290–299). In this respect, Ingarden might have been the only Polish philosopher on whom the influence of Brentano's deep religious charisma – many of Brentano's students, among them Husserl, converted to Protestantism under his influence – unexpectedly finds an echo.

Acknowledgments I am grateful to Sandra Lapointe for her comments and suggestions on a previous version of this paper.

References

Ajdukiewicz, Kazimierz (1948), 'Epistemologia i semiotyka' (Epistemology and Semiotics), *Przegląd Filozoficzny* 44, 336–347. Eng. trans. by J. Giedymin, *The Scientific World-Perspective and Other Essays 1931–1963*, Dordrecht, D. Reidel, 1968.

Ajdukiewicz, Kazimierz (1985a), 'Kierunki i prądy filozofii współczesnej' (Directions and Trends in Contemporary Philosophy), In *Język i poznanie* (*Language and Cognition*), vol. 1, Warsaw, PWN, 249–263.

Ajdukiewicz, Kazimierz (1985b), 'Postępowanie człowieka' (Human Action), In *Język i poznanie* (*Language and Cognition*), vol. 1, Warsaw, PWN, 317–364.

Ajdukiewicz, Kazimierz (2003), *Zagadnienia i kierunki filozofii* (*Problems and Theories of Philosophy*), Warsaw, Wydawnictwo Antyk-Fundacja Aletheia. Eng. trans. by H. Skolimowski and A. Quinton, Cambridge, Cambridge University Press, 1973.

Bolzano, Bernard (1837), *Wissenschaftslehre*, Sulzbach, Seidel.

Brentano, Franz (1930), 'Über den Begriff der Wahrheit', In *Wahrheit und Evidenz*, Oscar Kraus (ed.), Leipzig, Felix Meiner. Eng. trans. by R.M. Chisholm, *The True and the Evident*, London, Routledge and Kegan Paul, 1973.

Brentano, Franz (1933), *Kategorienlehre*, Hamburg, Felix Meiner. Eng. trans. by R.M. Chisholm and N. Guterman, *The Theory of Categories*, The Hague, Nijhoff, 1981.

Chrudzimski, Arkadiusz (1999), *Die Erkenntnistheorie von Roman Ingarden*, Dordrecht, Kluwer.

Czeżowski, Tadeusz (1948) *O metafizyce, jej kierunkach i zagadnieniach* (*On Metaphysics, Its Trends and Problems*), Toruń, T. Szczęsny.

Czeżowski, Tadeusz (1958), 'Zagadnienie istnienia świata w świetle przemian metodologicznych' (*The Problem of the Existence of the World on the Background of Methodological Changes*), In *Odczyty filozoficzne* (*Philosophical Lectures*), Toruń, Towarzystwo Naukowe w Toruniu, 34–43.

Czeżowski, Tadeusz (1958), 'Identyczność a indywiduum i jego trwanie' (*Identity and an Individual and Its Persistence*), In *Odczyty filozoficzne* (*Philosophical Lectures*), Toruń, Towarzystwo Naukowe w Toruniu, 160–169.

Ingarden, Roman (1951), 'Ingarden's Letters to T. Czeżowski', *Archiwum UMK, Spóścizna T. Czeżowskiego*, 185/V, 19 October 1951.

Ingarden, Roman (1964/65), *Der Streit um die Existenz der Welt*, Vol. I/II, Tübingen, Niemeyer.

Ingarden, Roman (1969), 'Le Concept de Philosophie chez Franz Brentano', *Archives de Philosophie* 32, 458–75.

Ingarden, Roman (1973), *The Literary Work of Art*, Eng. trans. by G. Grabowicz, Evanston, North-Western University Press.

Kotarbiński, Tadeusz (1913), 'Czy prawda jest tylko wieczna czy też i odwieczna?' (Is Truth Only Eternal or Both Eternal and Sempiternal), *Przegląd Filozoficzny* 16, 74–92.

Kotarbiński, Tadeusz (1961), *Elementy teorii poznania, logiki formalnej i metodologii nauk* (*Elements of the Theory of Knowledge, Formal Logic and Methodology of Sciences*), Wrocław, Wydawnictwo Ossolińskich. Eng. trans. by Olgierd Wojtasiewicz, *Gnosiology. The Scientific Approach to the Theory of Knowledge*, Oxford, Pergamon Press – Zakład Narodowy imienia Ossolińskich, 1966.

Küng, Guido (1963), *Ontologie und logistische Analyse der Sprache, Eine Untersuchung zur zeitgeönossischen Universaliendiskussion*, Vienna, Springer.

Leśniewski, Stanisław (1913), 'Czy prawda jest tylko wieczna czy też i odwieczna?' (Is Truth Only Eternal or Both Eternal and Sempiternal), *Nowe Tory* 18, 493–528. Eng. trans. by Leśniewski 1992.

Leśniewski, Stanisław (1992), 'On the Foundations of Mathematics', In *Stanisław Leśniewski,Collected Works*, vol. I, S.J. Surma, J.T. Srzednicki, D.I. Barnett and V. Rickey (eds.), Dordrecht, Kluwer.

Łukasiewicz, Dariusz (2002), *Filozofia Tadeusza Czeżowskiego* (*Tadeusz Czeżowski's Philosophy*), Bydgoszcz, Akademia Bydgoska.
Łukasiewicz, Dariusz (2006), 'Brentanian Philosophy and Czeżowski's Conception of Existence', In *Actions, Products and Things. Brentano and Polish Philosophy*, Arkadiusz Chrudzimski and Dariusz Łukasiewicz (eds.), Frankfurt, Ontos, 183–215.
Łukasiewicz, Dariusz (2007), 'Brentano's theory of judgment and the Lvov-Warsaw School', *Ruch Filozoficzny* 1, 33–47.
Przełęcki, Marian (1989), 'The Approach to Metaphysics in the Lvov-Warsaw School', In *The Vienna Circle and the Lvov –Warsaw School,* K. Szaniawski (ed.), Dordrecht, Kluwer, 55–66.
Smith, Barry (1990), 'On the Phases of Reism', In *Kotarbiński, Logic, Semantics and Ontology*, Jan Woleński (ed.), The Hague, Nijhoff, 137–185.
Twardowski, Kazimierz (1894), *Zur Lehre vom Inhalt und Gegenstand der Vorstellungen. Eine psychologische Untersuchung,* Vienna, Hölder; reprinted in *Philosophia*, Munich/Vienna, 1982. Eng. trans. by R. Grossmann, *On the Content and Object of Presentations*, The Hague, Martinus Nijhoff, 1977.
Twardowski, Kazimierz (1895), 'Metafizyka duszy' (*Metaphysics of the Soul*), *Przełom* 15, 467–480.
Twardowski, Kazimierz (1910), *O filozofii średniowiecznej wykładów sześć* (*Six Lectures on Medieval Philosophy*), Lvov.
Twardowski, Kazimierz (1921), Symbolomania i pragmatofobia (*Symbolomania and Pragmatophobia*), *Ruch Filozoficzny* 6, 1–10. Eng. trans. in Pelc 1978.
Twardowski, Kazimierz (1965), 'Przemówienie wygłoszone na zebraniu jubileuszowym Polskiego Towarzystwa Filozoficznego we Lwowie' (Address on the Occasion of the Jubilee Meeting of the Polish Philosophical Society in Lvov), Księga Pamiątkowa Polskiego Towarzystwa Filozoficznego we Lwowie (Memorial Book of the Polish Philosophical Society in Lvov), in *Wybrane Pisma Filozoficzne* (*Selected Philosophical Writings*), Warsaw, PWN, 379–384.
Woleński, Jan (1989), *Logic and Philosophy in the Lvov-Warsaw School*, Dordrecht, Kluwer.
Woleński, Jan (1990), 'Marty and the Lvov-Warsaw School', In *Mind, Meaning and Metaphysics. The Philosophy and Theory of Language of Anton Marty*, K. Mulligan (ed.), Dordrecht, Kluwer, 215–223.
Woleński, Jan (1990), 'Kotarbiński, Many-Valued Logic, and Truth', In *Kotarbiński, Logic Semantics and Ontology*, Jan Woleński (ed.), The Hague, Martinus Nijhoff, 191–197.
Woleński, Jan (1999a), 'Reism and Leśniewski's Ontology', In Woleński 1999b, 13–21.
Woodger H. Joseph (1939), 'The Technique of Theory Construction', *International Encyclopaedia of Unified Science*, vol. II, Chicago, University of Chicago Press.
Wundt Wilhelm (1906), Logik, Stuttgart.

Chapter 2
The Genesis and History of Twardowski's Theory of Actions and Products

Jerzy Bobryk

2.1 Introduction

Kazimierz Twardowski was born in 1866 in Vienna where he later studied and joined the circle of Brentano's followers, which included Edmund Husserl, Alexius Meinong, Karl Stumpf, and others. In 1892 he visited Wundt's psychological laboratory in Leipzig and studied shortly under Karl Stumpf in Halle. In 1894, his thesis *Zur Lehre vom Inhalt und Gegenstand der Vorstellungen* (*On the Content and Object of Presentation*) presented for the *Habilitation* qualified him as assistant professor. In 1895, he moved to Lvov (Lemberg) where he held a professorship at the Jan Kazimierz University for 35 years. Kazimierz Twardowski died in 1938.

Twardowski was the founder of the Lvov-Warsaw school, which is also sometimes referred to as the Polish school of Analytical Philosophy. Twardowski devoted his career to building in Poland a properly scientific philosophy. He also founded the first Polish laboratory of experimental psychology. He trained a number of famous Polish philosophers, logicians, sociologists and psychologists, like Władysław Tatarkiewicz, Tadeusz Kotarbiński, Kazimierz Ajdukiewicz, Innocenty Bocheński, Alfred Tarski, Jan Łukasiewicz, Stanisław Leśniewski, Stanisław Ossowski, Maria Ossowska, Władysław Witwicki, and others. Twardowski wrote in German and in Polish. His works deal with the analysis of philosophical, psychological, logical and ethical concepts.

It is not an easy task to give a brief summary of Twardowski's intellectual development. Roughly, his starting point was Brentano's theory of intentional acts, and although he never abandoned the basic assumptions of Brentano's philosophical psychology he ultimately reverted to the analytical method of the positivist tradition. In other words, Twardowski's philosophy was an original and clever mixture of phenomenology and linguistic philosophy and, in this respect, it is worth stressing that he carried out his project more than one hundred years ago. I would like to present

J. Bobryk (✉)
The Institute of Psychology of the Polish Academy of Science, The Institute of Philosophy at the Warsaw School of Social Psychology, Warsaw, Poland
e-mail: j.k.bobryk@wp.pl

S. Lapointe et al. (eds.), *The Golden Age of Polish Philosophy*, Logic, Epistemology, and the Unity of Science 16, DOI 10.1007/978-90-481-2401-5_3,

here only one aspect of the development of Twardowski's thought: his conception of intentional act and its evolution toward his theory of human action, that is, a theory dealing with the problem of the relationship between the human consciousness and human behaviour.

2.2 Brentano's Theory of Intentional Acts

Franz Brentano, in his *Psychologie vom empirischen Standpunkt* argues that mental phenomena have a distinctive property: they are intentional. Our mental states are, in some sense, directed at objects and states of affairs in the world. If someone has a belief, it is a belief that something is the case. If I have a wish, it must be a wish for something. If Mary or John are imagining, they must be imagining something. Our mental states may be directed at material or ideal or fictional objects (Bobryk 2001, Jadacki 1992). I can believe that the King of France is a handsome man even if there is no King of France. I can imagine Pegasus, I can think about complex numbers, I may dream of Beatrice. All these cases involve something that Brentano calls the "object" of consciousness and according to Brentano, every mental phenomenon is characterized by its intentional object. The intentional object of my act of perception is the perceived (material) object. The intentional object of my imagining Pegasus is the (fictional) mythological Pegasus. Pegasus can be an object of my believing. I believe that Pegasus exists in the same manner that I believe that the Eiffel Tower or the Empire State Building exist. "Directedness" or "aboutness" is the primary property of intentional acts.

2.3 Twardowski (1894/1977) on the Content and Object of Presentation

In *Zur Lehre vom Inhalt und Gegenstand der Vorstellungen*, Twardowski argued that Brentano's use of the word "object" was ambiguous: it may denote, on the one hand, an independently existing entity toward which our mental actions are directed, as it were, and, on the other, a mental product, a representation or presentation – the terminology is here indifferent – of an entity. What Twardowski has in mind is roughly the following. Imagine that Pegasus exists actually in the external world, say, as a result of genetic engineering. Were there to be a real, albeit artificial Pegasus, it seems that "object of the mental action" could mean two different things: the material Pegasus or its "mental representation". Twardowski called the former the transcendent object of the mental action, the latter the content or the immanent object of this action. The content is typically a mental or psychological entity, and it is the result of the mental processes of a particular agent. The transcendent object of a mental act is usually a material or physical entity and it is independent of the agent's mind.

A physically existing genetically-engineered Pegasus can be the object of my visual perception. According to Twardowski, the mental action of perception "produces" the mental representation of Pegasus. This "product", that is, the representation of Pegasus exists as long as the action of perceiving Pegasus lasts. A different mental action, say, the action of imagining or the action of remembering Pegasus, will produce an imagined or remembered Pegasus. The latter product, the imagined Pegasus, for its part, may exist even if the material Pegasus no longer exists. More generally, the objects about which I think, even when they are not material objects, like triangles or numbers, have objective properties. Any triangle has three straight sides and three angles even if I do not know that it does. In this sense, ideal or abstract objects are also mind-independent. (For more about Brentano and Twardowski's theories of intentionality, see also Dariusz Łukasiewicz, this volume, Bobryk 2001)

2.4 Acts of Consciousness and Psychophysical Actions

According to Twardowski, the difference between the content and the object of my (or your) action of imagining the Eiffel Tower is analogous to the difference between a camera picture of the Tower and the Tower itself. The content is a product of my mental act like a photo is the product of the photographic process. When the content of a mental act is the representation of a particular material object, then it is, according to Twardowski, in a very substantial way analogous to the product of a photographic process, when the latter is a representation of a material object. (Twardowski held a version of representationalism that will not be discussed here for the sake of brevity,) And just as we may produce the representation of non-existing object using, say, a digital camera and a computer, we may also imagine non-existing fictional objects. What is interesting about Twardowski's view is not his psychology *per se* but the way in which he connects psychology and semantics. According to him, when I am speaking about the Eiffel Tower, the latter is the content of both the mental act that underlies the utterance and of my speech act.

Twardowski saw an analogy between mental actions such as imagining or perceiving and certain types of "psychophysical" actions – actions that involve both our mind and our body – such as speaking. The analogy, as Twardowski understands it, consists in the fact that in both cases we can single out three elements: the action itself, its product or result, and its object or reference. What is also common to all three cases, according to Twardowski, is the fact that the object and the product of the action (apparently or factually) determine the course of the action. Twardowski might have conceived of this teleological determination as a special kind of causality that does not exists literally. Human actions are only "apparently" caused by their goals. But in the latter case, teleological determination is perhaps nothing more than a convenient way of describing human actions. I will not enter this discussion here.

2.5 Twardowski's Theory of Actions and Products

Consider the following pairs of words:

to believe – the belief,
to talk – the speech,
to walk – the walk,
to inform – the information.

In each pair, the first word denotes an action, the second the product of this action.[1] The difference between the two words is not only grammatical but also semantic since the two words have different denotation. When we argue, a discussion results. When we think, mental acts result. The nature of the relations between actions is an intricate metaphysical question which requires Twardowski to make a number of additional ontological distinctions. In particular, Twardowski draws an important distinction between *durable products* and *non-durable products*. The case in which a product almost merges with the action of which it is the result is different from the case in which a product is clearly distinct from the action. The first kind of product are typically non-durable: those product exists only as long as the action that produces them lasts. Speech exists only as long as the action of speaking continues, mental acts as long as someone is having beliefs and doubts or is otherwise being conscious. There are, however, products that can last longer than the action which produces them: the picture that results from my drawing, the text that results from my typing, or (although this is not Twardowski's example) the software that results from my programming the computer, for instance.

Twardowski also introduces an important new category when it comes to the characterization of actions and products:

> [A]mongst actions and products we may single out those that are psychophysical, namely those in which a physical action is accompanied by a mental action which somehow affects that physical action.
>
> (Twardowski 1912/1979, 15).

A text written on paper is usually the product of a human action that is simultaneously mental and physical and may therefore be called a psychophysical product. As a whole, what we call culture and civilization are indeed, according to Twardowski, a collection of such psychophysical products. Some of them – like books or paintings – are durable products; others – like musical performances – are non-durable. From the ontological point of view there are thus, according to Twardowski, three categories of actions and products: physical actions and products, mental or psychological actions and products, and psychophysical actions and products. In principle, there are six possible combinations of actions and products – one of which is empty – which are illustrated in the following table:

---

[1] Unless otherwise specified, what follows is based on Twardowski (1912/1979), see also Twardowski (1999); for an alternative analysis of the distinction and its significance, cf. also (Maria van der Schaar 2006).

Table 2.1 Actions and products

Actions and products	Non-durable	Durable
Physical	to dance, *the dance* to walk, *the walk*	to build, *the building* to cook, *the food*
Mental	to believe, *the belief* to perceive, *the perception*	
Psycho-physical	to speak, *the speech* to inform, *the information*	to paint, *the painting* to write, *the text*

Twardowski (1912/1979, 21) argues that a mental product is never durable by itself. It only ever exists as long as the mental action that produces it: it depends on the latter for its existence. The particular mental representation of the Eiffel Tower that results from my perceiving it exists in my mind as long as I actually perceive it. When the perception comes to an end, I may initiate a new mental action, say, that of imagining or remembering the previously perceived object, which brings about a new product, which is also ontologically dependent upon my performing the relevant action. Products of psychological processes are not permanent and they disappear when the process comes to an end. The fact that humans are often (wrongly) convinced of their durability is due to a certain confusion: mental actions are, by definition, involved in *psycho*physical actions and, if we follow Twardowski, we habitually produce and employ psychophysical products such as a written text, pictures, logical notations, etc., that is, as ways of, so to say, investing the fleeting products of human thinking with permanence.

Twardowski's thesis concerning the non-durability of mental products may seem controversial but it was vindicated by others, and in particular by the prominent cognitive psychologist Ulric Neisser – independently and over 50 years after Twardowski. The latter helps us throw light on Twardowski's own conception. According to Neisser (1967), traditional theories of human cognitive processes are based on the assumption that mental representations are durable. Neisser termed this presupposition "the reappearance hypothesis", which, in his opinion, originated in English Empiricism. Classical psychology assumes that cognitive units like "ideas" or "concepts" are stored in the human mind and can be reused several times. If we follow the latter, the same memory, image, idea, etc., can disappear and reappear to an agent over and over again (Neisser 1967, 281). To the reappearance hypothesis, Neisser opposed his own view, "the utilization hypothesis". The latter is based on the assumption that the products of mental activity are, as Twardowski believed, transient. According to Neisser, cognitive units are not fixed products separate from the acts themselves and, which, once created, continue to exist in a ready form and may be reused many times. The human mind stores not so much ready ideas, concepts and representations, as traces of previous cognitive acts. What is repeated is the act itself, and repeated cognitive activity brings about similar results.

> Recall by way of an image takes places when a new construction is largely under the control of what remains from an earlier one.
>
> (Neisser 1967, 285).

According to Neisser, traces of cognitive acts are not simply "revived" or "reactivated" in the act of recall. Instead, the stored fragments are used as information to support a new "construction".

> It is as if the bone fragments used by the palaeontologist did not appear in the model he builds at all – as indeed they need not, if it is to represent a fully fleshed-out, skin-covered dinosaur. The bones can be thought of, somewhat loosely, as remnants of the structure which created and supported the original dinosaur, and thus as a source of information about how to reconstruct it.
>
> (Neisser 1967, 286).

2.6 Mental Acts and Brain Processes

Twardowski's investigation of mental acts and events naturally led him to ponder on the mind-body problem, that is, the problem of explaining how human mental states and acts are related to the states and processes in our brain. Can we assume that the words "mind" and "brain" are two names of the same thing? Since the seventeenth century philosophers had remained convinced that mental phenomena are "subjective" whereas (physiological) brain processes are "objective" and Twardowski, at least at first glance, seems to follow his predecessors. According to Twardowski while one can study the human mind through psychological (for instance, introspective) methods, the brain ought to be studied on the basis of an objective experimental methodology. But does this division of labour imply the non-identity of the mind and the brain?

There are two ways of conceiving of the mind-body problem according to Twardowski (1897/1997, 90), which derive from two ways of understanding the idea that the mind is a function of the brain. The term "function" – which is involved in his analysis of the problem has (in both Polish and English) at least two meanings. The first one is mathematical or quasi-mathematical: when we say that *X* is a function of *Y*, we mean that a variable quantity *X* depends on variable quantity *Y*. For example, a meteorologist would say that air pressure is a function of temperature because air pressure changes when air temperature changes. On the second (Aristotelian) understanding, the term "function" designates the characteristic activity or purpose of a thing or person. In the later sense, we say that the function of the heart is to pump blood through the body or that the function of the teacher is to teach students. Psychological and philosophical discourse is not immune to this ambiguity. Hence, with respect to the idea that the mind is a function of the brain – that mental processes are a function of brain processes – we may mean, on the one hand, that mental states and processes correlate with changes in the brain states and processes. This would be, according to Twardowski (1897/1997, 90–91), the quasi-mathematical understanding of the expression. According to Twardowski the sentence: "The mind is a function of the brain" is entirely and undoubtedly true when taken in the quasi-mathematical sense: our mental states have to change when our brain processes change. In this, he could plausibly be seen to anticipate a popular contemporary position according to which:

> Mental phenomena, all mental phenomena whether conscious or unconscious, visual or auditory, pains, tickles, itches, thoughts, indeed, all of our mental life, are caused by processes going on in the brain.
>
> (Searle 1984, 18)

However, the same expression may be understood, on the other hand, as meaning that the function of the brain is to perform various mental acts or processes, like thinking, feeling, or being aware of something. According to Twardowski, the sentence "the mind is a function of the brain" is false if we take the word "function" as implying some kind of purposeful activity. And the reason for this is that the brain on its own is unable to perform mental acts. According to Twardowski, mental acts, just like physical acts, are performed *by the whole person* (Twardowski 1897/1997, 90–91). If the human brain was isolated from the rest of the body it could not perform mental actions. We can explain what Twardowski has in mind in the following way. Let us imagine a science-fiction scenario in which, as a result of some surgical procedure, the brain has been removed from the body and placed on some kind of life-support system. I believe, and Twardowski would believe, that a brain isolated from the rest of the body, that is, from the totality of its input system as well as the entirety of its output mechanism (the configuration of muscles, bone, skin, etc. that constitute the body parts it controls) and placed on life-support could not perform its normal psychological functions. Note that the situation Twardowski has in mind is different from the one Putnam has in mind in the "brain in a vat" thought-experiment – although Twardowski's conclusions are not incompatible with the latter. Twardowski's scenario is not meant as a thought experiment but as an actual description of what is the case: while neurophysiological processes are performed by the brain (or in the brain), normal psychological functions such as seeing and remembering are performed by the whole person (at least in part, one could add, because they almost always involve kinaesthetic experience). If Twardowski is right, then the thesis according to which mental processes and brain processes are identical must be rejected.

According to Twardowski, products of mental actions are not durable. Human beings may however perform psychophysical actions that result in durable products. These psychophysical products become, so to say, external manifestations of non-durable mental products. A psychophysical product (durable or not) is the manifestation of a particular mental product only under the following condition: because it is apt to evoke a mental action which is identical or similar to the action from which it initially resulted, the psychophysical product (in which the mental product is manifested) can itself become the cause of the emergence of another identical or similar mental product. Thus, according to Twardowski, an impressionist painting may cause the viewer to have a mental state at least similar to one of the mental states in which the artist found himself while he was performing the action of painting.

Twardowski also calls "signs" (of mental products) durable psychophysical products that "manifest" mental products in this sense. Note that "sign" here should *not* be understood to be restricted to linguistic signs but to any durable psychophysical products which can exist independently from the actions of a human subject and which have a potential "meaning". Some durable psychophysical products have the capacity to (partially) cause mental actions in the mind of she who perceive them that are identical or in some significant way similar to those that took place in the mind of the author and hence result in similar products.

According to Twardowski, there are two types of sign use. In the first case, ongoing cognitive acts bring about psychophysical products that are symbols or expressions for a "meaning" in that particular context (that is, as long as the psychophysical action continues). In the second case, the symbol is separated from the action that produced it and subsists by itself. This symbol may in turn cause other cognitive acts. Note here that Twardowski implicitly subscribes to Neisser's idea that what is recalled or evoked are acts, not their content. A sign may evoke a range of possible cognitive actions – in the same or in another subject. According to Twardowski, there are as many different products as there are actions that may possibly be evoked to produce them. (Twardowski 1912/1979, 22, Bobryk 2001, 38) Such products are not therefore identical, but differ from one another to a greater or lesser extent – Twardowski supposes here that contents are individuated by the action that produce them. Nonetheless, these various mental products have certain common characteristics. The latter are elements that individual mental products share and are what we usually understand to be their (objective) meaning or content.

Amongst psychophysical products there are some which Twardowski calls "artefacts", who stand for or act as surrogates for other products. (Twardowski 1912/1979, 23, Bobryk 2001, 56). Twardowski uses this idea to explain, for instance, what happens in atypical and, in particular, "pretense" situations of linguistic exchange. When an actor assumes a posture to express a certain emotion, it is understood that his posture merely "depicts" this emotion, that it is not a result of a genuine emotion. The same could be said about the "words" uttered randomly by a parrot.

More interestingly, according to Twardowski, logic also makes use of artefacts or surrogate signs. (1912/1979, 23) A proposition, according to Twardowski, is the product of an action of believing – it can be seen to make up its content – and it is what is typically expressed in the psychophysical products which result from the psychophysical action of uttering or making a statement. According to Twardowski, statements thus express propositions, so that propositions are the meaning of statements. (Twardowski 1912/1979, 24). His view on the nature of logical formulae however – the latter are not mere propositions – consists in saying that:

> we can, however, produce artificial, surrogate statements which are not expressions of propositions actually made, but expressions of artificial products that are surrogate for propositions actually made.
>
> (Twardowski 1912/1979, 24)

We use surrogate products (artefacts) in science as well as in everyday life and very often genuine and surrogate products occur alternately. As Twardowski sees it, the complex cognitive actions involved in scientific knowledge are essentially an interplay of propositions and concepts whose meaning has been developed in the mind of the agent, and of abbreviated, surrogate representations (imaginations of abstract symbols) of other propositions and concepts, whose contents are not developed in order to facilitate their manipulation. Cognitive agents are thus in a position to use the products of psychological actions independently of the psychological actions themselves, and to do so without having to repeat fully the process that led to the creation of those products. The symbolism of formal logic, for instance, enables the use of "records" for propositions which in fact do not result from acts of judgement but merely, so to say, stand in for the latter. The use of such symbols – in the sense of surrogate psychophysical products – instead of substantively developed propositions simplifies and standardizes cognitive operations and, at the same time, renders them more foolproof under typical conditions. The possibility of a partial or a total separation of products of mental actions from the actions themselves implies that signs can be used independently from psychological meanings, at least independently of those of their producer.

The ordinary and banal character of these phenomena (the possibility of a mechanical use of the signs of a semiotic system once it has been invented) often makes it hard to assess correctly the role these thought-recording systems may play in cognitive activity. Written and formal languages are as many different ways of availing oneself of the psychophysical products while keeping them separate from the cognitive actions themselves. Twardowski, one may speculate, would say the same about the nature of computer software (Bobryk 1989). Yet, Twardowski would obviously refuse to think of computers as models of the human mind. The computer is no more a proper model of the human mind, than my foot is a model of my locomotive system (albeit that it is a part of this system). Moreover, computer information processing is no more a simulation of my thinking than my cycling is a simulation of my running, albeit that in both case I can obtain the same result: I can move from point A to point B.

References

Bobryk, Jerzy (1989), 'Cognitive science – the science of artefacts', *Polish Psychological Bulletin*, vol. 20, no 1, 3–14.

Bobryk, Jerzy (2001), *Twardowski. Teoria działania (Twardowski. Theory of action)*, Warszawa: Prószyński I S-ka.

Jadacki Jacek J. (1992), 'The metaphysical basis of Kazimierz Twardowski's descriptive semiotics', In Jacek Paśniczek (ed.), *Theories of Objects: Meinong and Twardowski.* Lublin: Wydawnictwa UMCS.

Neisser, Ulric (1967), *Cognitive Psychology.* New York: Appelton-Century-Crofts.

Schaar, Maria van der (2006), 'On the ambiguities of the term judgement. An evaluation of Twardowskis distinction between action and product', In *Actions, Products and Things; Brentano and Polish Philosophy*, A. Chrudzimski and D. Lukasiewicz (eds.), Frankfurt: Ontos Verlag, 35–53.

Searle, John (1984), *Minds, Brains and Science.* Cambridge, Massachusetts: Harvard University Press.

Twardowski, Kazimierz (1894), *On the Content and Object of Presentation.* The Hague: Martinus Nijhoff, 1977.
Twardowski, Kazimierz (1912), 'Actions and products. Comments on the border area of psychology, grammar and logic', In Jerzy Pelc (ed.), *Semiotics in Poland 1894–1969*, Boston: Reidel, 1979. *Kazimierz Twardowski on Actions, Products and Other Topics in Philosophy*, Arthur Szylewicz, Johannes Brandl, and Jan Wolenski (eds.), Amsterdam: Rodopi, 1999.
Twardowski, Kazimierz (1897), 'Psychologia wobec fizjologii i filozofii' (Psychology in the face of physiology and philosophy), In Teresa Rzepa (ed.), *Psychologia w szkole lwowsko-warszawskiej (Psychology of the Lvov-Warsaw School)*, Warszawa: PWN, 1997.

Chapter 3
The Rise and Development of Logical Semantics in Poland

Jan Woleński

3.1 Introduction

The aim of this paper is to portray the rise and development of logical (formal) semantics in Poland in the years that constitute the "golden age" of Polish logic and philosophy (see also Woleński 2002). I use the adjective "Polish" purely geographically, that is, as equivalent to "in Poland". This means that the phrase does not connote any national ingredients. My considerations are restricted to the period that extends from 1895 (when Kazimierz Twardowski came to Lvov as the professor of philosophy) until 1939 (the beginning of World War II), but I'll make some comments about the situation after 1945. Technically speaking, there was no Poland between 1895–1918, but Polish philosophy nonetheless developed quite well during those years, and I will abstract from these peculiar political circumstance. Since the length of this paper is limited, my presentation focuses on certain issues and ignores others. In particular, I do not compare Polish philosophy with other contemporary traditions or schools in which logic and semantics flourished. I also assume that the ideas of the main Polish philosophers, especially those of Tarski, are sufficiently well known and do not require a detailed presentation.

3.2 Twardowski and his Role

Although semantics arose in Poland as a result of the cooperation between mathematical logicians and philosophers (see also Murawski, in this volume), its origin can be traced back to the work and activities of one man, Kazimierz Twardowski. Twardowski wanted to implant the philosophical ideas of his teacher Franz Brentano in Poland. Brentano promoted clear and precise philosophy and, for the task, he considered language to be a very important tool. As a matter of fact, Brentano had himself proposed a reform of logic and although the latter as such had little impact, some of his ideas (as well as ideas of his students', in particular, Edmund

J. Woleński (✉)
Jagiellonian University, Kraków, Poland
e-mail: wolenski@if.uj.edu.pl

S. Lapointe et al. (eds.), *The Golden Age of Polish Philosophy*, Logic, Epistemology, and the Unity of Science 16, DOI 10.1007/978-90-481-2401-5_4,

Husserl, Anton Marty and Alexius Meinong) were significant for the development of philosophy of language as a whole. One can mention, for instance, the view that propositions are not combinations of concepts (the idiogenic theory of judgments) and the distinction between "synsemantica" (syncategoremic expressions, like "and", "or", etc.) and "categoremata" (categorematic expressions, like sentences and names). Moreover, in his later work Brentano developed a position known as reism, that is, a view that implies that abstracta (universals) exists neither in language nor in reality. On this latter view, abstract nouns are considered to be synsemantica.

Twardowski is at the source of the intense Polish interests in semantics. Before him, remarks on semantics in logic textbooks by Polish authors were very elementary. Twardowski, in addition to his programmatic insistence on the importance of being clear when doing philosophy, proposed two significant doctrines. First, he suggested that there is a close connection and interplay between descriptive psychology and semantics (or logic in his sense):

> [...] if a name really yields an accurate picture of the mental state of affairs which corresponds to it, then it must also show a correlate to the act of presentation. Indeed, there is such a correlate; and to the three aspects of presentation – the act, the content, and the object – there corresponds threefold task which every name has to fulfill.
>
> (Twardowski 1894, 8–9)

The three tasks names are meant to fulfill, according to Twardowski, consist in (i) awakening a related mental content in the listener, (ii) signifying the existence of the corresponding mental content (the "meaning" of the name in question; quotes in the original) in the speaker, and (iii) designating objects. Proposing that names fulfill these three tasks was not a novelty *per se*. The actual importance of Twardowski's remark should be seen as twofold. Firstly, he clearly derives the referential import of names from the intentional character of mental acts. Secondly, Twardowski maintains that the properties of presentations can, at least to some extent, be established through an analysis of names and their functions. One example of this is Twardowski's discussion of the relatively famous problem of objectless presentations (see Twardowski 1894, 20–21). Are there concepts that have no object – are there "objectless presentations"? Twardowski's answer, suprisingly, was negative. One famous argument, at least among Brentano's students, for the existence of objectless presentations consisted in saying roughly the following: we have the concept of nothing and because this concept refers precisely to nothing (that is, to no object), it must be objectless. Twardowski's reply rests on a logical analysis of the word "nothing". Twardowski first considers what is usually called the *infinitatio* (or dual complement) of a term ("non-man", for instance, is the *infinitatio* of "man"). He then observes that if *N* is a name and *non-N* its *infinitatio*, there must be a name *N'* whose scope is *N* + *no-N*. Conversely, every general name can be "divided" into two others, which are complementary so that both together exhaust the universal extension. Yet, there is no category which corresponds to the union of "nothing" and "something" (the putative dual complement of the latter). Hence, according to Twardowski, "nothing" is not a name. Sentences containing the term "nothing" must be reformulated or paraphrased entirely. For instance, "nothing is eternal" means that "There is something which is eternal" must be negated or is false. In contemporary terms, this amounts to saying that "nothing is eternal" means that there is no *x*, such that *x* is eternal. What this implies is that, in fact, for Twardowski "nothing" stands for a quantificational expression. Twardowski's analysis of "nothing" was the first attempt in Poland to analyze philosophical problems via logical analysis.

Twardowski's second important contribution to semantics rests in his distinction between "actions" and "products". (see Twardowski 1912 and Bobryk, this volume) If we follow Twardowski, it is in principle possible to associate each human action with its outcome – what Twardowski calls its "product" – in such a way that these associations can then be expressed by pairs of the form <verb, associated noun>, for example: <to sing, song>, <to paint, picture>. In particular, there are the *psychophysical actions* and their products. Language is of the latter kind. Linguistic acts are psychophysical. They necessarily consist in manipulating some physical objects (symbols, sounds) but they do not consist only in that. In order for there being an act of language, there also has to be an agent who performs this acts and, while doing so, this agent typically engages in some mental processing which, in turn, "manifests" itself as the meaning of his expression. Let us leave aside the question of explaining what "manifest" means in this context and what happens in some non typical cases of communication (such as theater, for instance) or pathological ones (such as the case where the agent produces sound-patterns perfunctorily). According to Twardowski, these meanings are liable to being "objectified", what amounts to saying that although meaning does not exist independently of actions and their products, it supervenes in a sense on real objects. In this way, Twardowski hoped to avoid the dilemmas between Platonism and nominalism, on the one hand, and psychologism and antipsychologism, on the other.

3.3 How Semantics Was Done in Poland: Other Early Examples (1910–1920)

Twardowski was not a logician and his semantic ideas were not particularly influenced by logic. On the other hand, many of his students were engaged in sophisticated logical semantics. In the decade 1910–1920, Jan Łukasiewicz, Stanisław Leśniewski and Tadeusz Czeżowski were all pursuing investigations of the latter type. In general, three factors helped shape the thought of Twardowski's students, namely (i) Frege's influence, (ii) the problem of the antinomies and (iii) the rise of many-valued logic.

Frege was indeed fairly popular in Poland during this period (see Woleński 2004). Czeżowski (see Czeżowski 1918) and Łukasiewicz (see Łukasiewicz 1920) adopted different aspects of his semantics and in particular the view that the True and the False are the referents of sentences. Łukasiewicz was unmistakably clear on this (and it must be noted that he later on came to reject this view):

> [...] by *truth* I mean not a true proposition, but the *object* denoted by a true proposition, and by *falsehood* I mean not a false proposition, but the *object* denoted by a false proposition [...] Two different true propositions, for instance "*2* × *2* is *4*" and "Warsaw lies on the Vistula" differ only by their *contents*, but they denote *the same* object, that is truth, in the same way as the expressions "*2* × *2*" and "*3* + *1*" differ only by their contents, but denote *the same* object, that is the number *4*. All true propositions denote one and the same object, namely truth, and false propositions denote one and the same object, namely falsehood. I consider truth and falsehood to be *singular* objects in the same sense as the number *2* or *4* is. There are as many different names of the one and only truth as there are true propositions, and as many different names of the one and only falsehood as there are false propositions. Ontologically, truth has its analogue in being, and falsehood, in non-being. The objects denoted by propositions are called *logical values*. (1920, 90)

Although the case is not as clear as far as Łukasiewicz is concerned, it can also be documented. For Łukasiewicz, as for Frege, truth and falsehood are not defined and their explanations have an intuitive character. In Łukasiewicz's eye, this approach however justifies the definition of logic as the science of logical values and he evokes Frege's account of logic as the science of truth. Because he viewed logic as the science of logical values, Łukasiewicz felt justified to use the rules governing 1 and 0 as the basis for the deduction of logical principles. However, in this respect, Łukasiewicz goes beyond Frege who did not use truth and falsehood as logical constants. It could be argued that Łukasiewicz's idea that possibility is the third value for sentences was motivated by the Fregean-approach to logical values. (On Łukasiewicz's three-valued logic, see Malinowski, this volume.) At any rate, (Łukasiewicz 1920) was meant to be the introduction to a extensive work on many-valued logic and although the latter was never written, Łukasiewicz's three-valued logic was the first formal semantics to go beyond bivalence.

Łukasiewicz also worked on antinomies and, in particular, on the liar paradox. He introduced a canonic way of presenting this antinomy:

> There are mental constructions which seem to contain an inevitable contradiction. For example, the sentence: line 13 on p. XXXV of this book [in which Łukasiewicz' paper was included – J. W.] contains a false sentence, is a construction of this kind. This sentence contains a contradiction, because observing that this sentence contains itself in the line 13 on p. XXXV of this book, it is easy to prove that its falsity entails its truth.

Łukasiewicz also proposed a solution to the paradoxes:

> [...] every logical principle contains *variables*. [...] These variables, like variables in mathematics, can have various values. Now there is a logical law which says that all logical principles concern only *those* objects that are *values of variables*. One can show that the above sentence containing the contradiction cannot be the value of a variable. Hence, logical principles do not apply to this sentence; this construction is outside logic.
>
> (Łukasiewicz 1915, p. xxxv)

Thus, according to Łukasiewicz, the liar paradox arises because the liar sentence violates the principles of correct logical languages.

The early Leśniewski also dealt with antinomies. I will not discuss the solution he attempts in (Leśniewski 1913), because it would require a lengthy prologue (see Betti 2004 for details). What is relevant and interesting for our purpose concerns Leśniewski's approach to language:

> Since keeping to 'natural intuitions' of language we get involved in irresolvable paradoxes, these 'intuitions' seem to imply contradiction. The '*artificial*' frame of strict conventions is thus a far better instrument of reason than the language dissolving in the opaque contours of '*natural*' habits which often imply incurable contradictions – much as 'artificiality' regulated Panama Canal is a better waterway than the 'natural' rapids on the Dnieper.
> (Leśniewski 1913, 82)

Although Leśniewski did not use formal languages in his early work, the last quotation makes the message clear: artificial languages are much better tools for doing logic and semantics than ordinary speech.

3.4 Semantics in Poland 1920–1939. General Remarks

Polish linguists, logicians and philosophers, like their colleagues abroad, did not use the term "semantics" (or its Polish counterpart "*semantyka*") very frequently before 1930. We find in (Szober 1923) – one of the chapters is called "Semantyka" – what may be the earliest use of the word. The author, a distinguished Polish linguist, understood this term in the standard way, that is, as signifying the science of meaning; Szober uses the word in the same way in (Szober 1924). The word *semantyka* occurs in the title of (Niedźwiedzka-Ossowska 1925), incidentally an extensive review of Szober 1924. Stanisław Ossowski (see Ossowski 1926) spoke about semantic entities, that is, objects denoting, representing or having meaning. In Kotarbiński (1929, 15), semantics is understood as dealing with the meaning of linguistic expressions. Leśniewski introduced the label "semantic categories" ("*kategorie semantyczne*" in Polish) for Husserl's *Bedeutungskategorien*:

> In 1922 I outlined a concept of semantical categories as a replacement for the hierarchy of types, which is quite unintuitive to me. Frankly, I would still feel obliged to accept this concept even if there were no antinomies at all. From a formal point of view, my concept of semantic categories is closely related to the well-known theory of types [...] especially for their theoretical consequences. Intuitively, however, the concept is more easily related to the thread of tradition running through Aristotle's categories, the parts of speech of traditional grammar, and Husserl's meaning categories. (1929, 421)

Kazimierz Ajdukiewicz, another student of Twardowski, employed the word *semantyka* in his review (see Ajdukiewicz 1930) of (Kotarbiński 1929). The following year (see Ajdukiewicz 1931), Ajdukiewicz discussed semantic functions and, in particular, meaning and he delivered a course on "logical semantics" at the University of Lvov. It appears to have been the first occurrence of "logical semantics" (*semantyka logiczna*) on the cursus. This course (see Ajdukiewicz 1993) was devoted to the theory of semantic categories and the antinomies. In fact, Ajdukiewicz discussed many problems of logical syntax. The idea that semantics should be viewed as logical syntax was also proposed – independently – by Leon Chwistek, who developed (see Chwistek 1935) a nominalistic theory of linguistic signs as the foundations of logic and mathematics.

Alfred Tarski, in one of his first papers about the semantic conception of truth (see Tarski 1930) used the adjective *hererosemantyczny* ("heretosemantic", the word is taken from Leśniewski; "hererological" is standard today). (Tarski 1932) contains a reference to *Semasiologie* and a remark to the effect that the concept of truth is of "semasiological character". He adds that various results concerning the notion of truth can be extended to other semasiological concepts. More importantly, Tarski says that whenever we define semasiological concepts like that of truth for a language *L*, we must do so in its metalanguage *ML*. He also expressed doubts as to whether natural language is suitable for semasiological investigations. Tarski mentions satisfaction as another important semasiological concept. Although Tarski did not say this explicitly at this stage – he will shortly thereafter – we can indirectly conclude from his remarks that the fundamental concern of semasiology, according to his project, consists in investigating the relations holding between, on the one hand, linguistic expressions – sentences in particular – and, on the other hand, the objects to which the latter refer.

Tarski's very condensed remarks on semantics/semasiology are developed at length in his works throughout the 1930s, in particular in (Tarski 1933) and (Tarski 1936). The nature of these developments is illustrated in the following passages:

(a) [. . .] we attempted to go further and to construct [. . .] definitions and concepts belonging to the semantics of a language – i. e. such concepts as satisfaction, denoting, truth, definability, and so on. A characteristic feature of semantic concepts is that they give expression to certain relations between the expressions of language and the objects about which these expressions speak, or that by means of such relations they characterize certain classes of expressions or other objects. We could also say (making use of the *suppositio materialis*) that concepts serve to set up the correlation between names of expressions and the expressions themselves.

(Tarski 1933, 252)

(b) The word 'semantics' is used here in a narrower sense than usual. We shall understand by semantics the considerations concerning those concepts which, roughly speaking, express certain connexions between expressions of a language and the objects and states of affairs referred to by these expressions. As typical of semantical concepts we can mention the concepts of *denotation*, *satisfaction*, and *definition*, [. . .]. At least in its classical interpretation, the concept of *truth*– and this is not commonly recognized – is also to be included here.

(Tarski 1936, 401)

The problem of explaining how the concept of meaning is related to semantics affords its share of trouble for philosophers. Yet, it would be difficult to imagine how philosophy could ignore one of the main problem that comes along with the fact that expressions mean something, namely the problem of knowing what they mean. How does semantics accommodate the concept of meaning? If we assume that the classical tripartite division of semantics *sensu largo* is exhaustive, there are indeed only three possibilities: (a) the concept of meaning is to be dealt with as a part of syntax; (b) the concept of meaning is to be dealt with as a part of semantics *sensu sricto*; (c) the concept of meaning is to be dealt with as a part of pragmatics. Logical empiricists opted for (a), as is well known, without success. Solution (b) requires either that the concept of meaning be viewed as a new semantic primitive or that it be reduced to the concept of reference. It seems that augmenting the semantic vocabulary with the concept of meaning introduces a schism within semantics in the narrow sense, which is neither elegant nor easy to explain. Tarski himself did not address this problem directly. His general nominalistic sympathies made him, however, suspicious of the concept of meaning. He tried to find a solution by working with interpreted languages:

> It remains perhaps to add that we are not interested here in 'formal' languages for sciences in any special sense of the word 'formal', namely sciences in which no material sense is attached to the signs and expressions. For such sciences the problem here discussed [the problem of truth – J. W.] has no relevance, it is not even meaningful. We shall always ascribe quite concrete and, for us, intelligible meanings to the signs which occur in the language we shall consider. The expressions which we call sentences still remain sentences after the signs which occur in them have been translated into colloquial language. The sentences which are distinguished as axioms seem to us to be materially true, and in choosing rules of inference we are always guided by the principle that when such rules are applied to true sentences the sentences obtained by their use should also be true.
>
> (Tarski 1933, 166–167)

In particular, for Tarski, there is no inherent contradiction in the idea of a language which is simultaneously formal and interpreted. Although Tarski did not explain what it means for an expression to have "intelligible meaning", it is clear that an understanding of signs is a *condition sine qua non* for doing semantics (On this problem, see Patterson, this volume). This view is absolutely crucial for comprehending semantics as it was done in Poland. The view that Tarskian semantics is possible only for purely formal and artificial languages is simply mistaken. Of course, Tarski's views leave many points open to further discussion. In particular, we can wonder whether ordinary language is amenable to exact semantic analyses. As we know, according to Tarski, there are various limitations in this respect, but the claim that every meaningful expression is translatable into an ordinary expression is interesting in itself. Furthermore, in making interpreted languages the subject of semantics, Tarski thought that he could neutralize the distinction between propositions and sentences. Since interpreted sentences have a meaning we may, he thought, at least in formal semantics, dispense with propositions.

In many ways, Tarski's views were a continuation of Leśniewski's:

> Having no predilection for 'various mathematical games' that consist in writing out according to one or another conventional rule various more or less picturesque formulae which need not be meaningful or even – as some of the 'mathematical gamers' might prefer – which should necessarily be meaningless, I would not have taken the trouble to systematize and to often check quite scrupulously the directives of my system, had I not imputed to its theses a certain specific and completely determined sense, in virtue of which its axioms, definitions and final directives [. . .] have for me an irresistible intuitive validity. I see no contradiction therefore, in saying that I advocate a rather radical 'formalism' in the construction of my system even though I am an obdurate 'intuitionist'. Having endeavoured to express my thoughts on various particular topics by representing them as a series of propositions meaningful in various deductive theories, and to derive one proposition from others in a way that would harmonize with the way I finally considered intuitively binding, I know no method more effective for acquainting the reader with my logical intuitions than the method of formalizing any deductive theory to be set forth. By no means do theories under the influence of such formalizations cease to consists of genuinely meaningful propositions

> which for me are intuitively valid. But I always view the method of carrying out mathematical deduction on an 'intuitionistic' basis of various logical secrets as considerably less expedient method.
>
> (Leśniewski 1929, 487)

I think that the adjective "intuitionistic" should be replaced in the above quotation by "intutive". At any rate, Leśniewskian-type "intuitive formalism" shaped the general philosophical framework for semantics in Poland. This view is very closely related to Twardowski's distinction between actions and products. In fact, Leśniewski and Tarski both considered language to arise from acts of using expressions.

Intuitive formalism suggests that pragmatics may have priority over semantics, but this has little if any implication as regards the question whether the concept of meaning is subject to exact logical analysis. On the other hand, Tarski believed that syntactic forms and referential relations should be considered mathematically. In fact, according to Tarski, we should assume a parallelism between syntax and semantics which is manifested in the principle of compositionality. When we examine the Tarskian definition of a correct formula and the conditions under which compound expressions are satisfied or not, we see that everything proceed inductively from simpler to more complex cases. At the semantic level, compositionality concerns referential relations, not meanings, whatever they are. The principle, in other words, favours extensional contexts over intensional ones. Thus, the principle of extensionality is a by-product of the principle of compositionality. In fact, Polish logicians and philosophers considered intensional contexts, even those with alethic modalities, to be logically defective. Hence the several attempts to "extensionalize" intensionalities. This also explains why many-valued logic was considered to be the refuge of modalities: many-valued logic is extensional, that is, all logical functions occurring in it are fully extensional.

I already mentioned the *L/ML* distinction. The latter possesses a crucial significance for semantics. The reason for this rests not only in the fact that semantic theory is formulated in *ML*, but also in the fact that the distinction in question is closely associated with the requirement that a semantic metalanguage satisfy appropriate conditions. This is obvious when we construct the semantic theory of truth. If we construct the theory of truth for a language *L*, the metalanguage *ML* must be stronger than the object language. The theory of semantic (or syntactic) categories, formulated by Leśniewski and developed by Ajdukiewicz (see Ajdukiewicz 1936; the latter paper inaugurated categorical grammar), when combined to the *L/ML* distinction suggests quite naturally that the metalanguage must have expressions of a higher category than *L* itself. According to Leśniewski (see above), the concept of semantic category is a (more natural) substitute for Russell's notion of a logical type. In fact, the theory of semantic categories can be considered to be equivalent to the simple theory of logical types. Once we have the *L/ML* distinction and the theory of syntactic categories, we have resources rich enough to develop semantics for arbitrary languages. The full import of these results was not however properly assessed before 1945. Since, at least for the purpose of metamathematical investigations, every theory can be elementarized, that is, formulated in a first-order language, the semantic theory for any formalism can be given in a second-order logic. For example, if T is a first-order theory, a truth-definition for it must be done in a weak second order logic with arithmetical comprehension – although some amount of set theory is always required for doing semantics, it need not be excessively strong.

3.5 Tarski's Contributions and Their Importance

It is difficult to overestimate the import of Tarski's contributions to semantics. He should be considered to have been the main architect of the semantic revolution in logic (see Woleński 1999, for a report). His semantic theory of truth is perhaps the most important singular achievement of semantics and logic in Poland. Its significance consists not only in having provided a definition of truth, a diagnosis for semantic paradoxes as well as a method to spell out other semantic concepts, but also in having shown the limitations to which a truth-definition is subject. It should be mentioned that some of the details of the semantic truth-definition and the semantic paradoxes were, however, first elaborated by Leśniewski, Tarski's teacher. Furthermore, while Tarski (see Tarski 1936a) was the first to give an exact definition of logical consequence (entailment) – in contemporary terms: a sentence A is a logical consequence of a set X of sentences if and only if every model of X is also a model of A – this definition takes over significant aspects of Bolzano's definition of deducibility (*Ableitbarkeit*). In a way, it is through Tarski's remarkable achievement that the greatness of Bolzano as a logician could at last be properly assessed (see Siebel 1996, Chapter 5 for a comparison of Bolzano and Tarski).

Tarski's undefinability theorem says that the set of true propositions of a theory T sufficiently strong to include Peano arithmetic is not arithmetically definable. In other terms, if T includes Peano arithmetic, there is no formula of T which can be considered as defining the set of truths of T. This theorem, together with Gödel's incompleteness theorems, the Church theorem and the Löwenheim-Skolem theorem, belong to the so-called "limitative" theorems, which exhibit the inherent limitations of rich formal systems. But Tarski (see Tarski 1939) developed a general semantic method for proving undecidability, undefinability and incompleteness. These results throw light on the relation between syntax and semantics. Generally speaking, Hilbert's program and the "syntacticism" of the Vienna Circle proposed a reduction of all questions concerning formal systems to matters of syntax. Tarski's results (the same is suggested by Gödel's incompleteness theorems, although less directly) showed that syntax is too poor to cover semantics. Even if we say that higher-order syntax is enough for lower-order semantics, we beg the question in a sense, because the former is in its very essence semantics. Ultimately, higher-order syntax is always a part of set theory and set theory is, one could claim, not only the foundations of mathematics, but also the base for semantics. Tarski's semantic works resulted in model theory. Although this chapter belongs to the years after 1945, it should be mentioned. After the new approach to model theory based on universal algebra were developed, the study of formal systems became a legitimate part of metamathematics, next to proof theory and recursion theory.

3.6 Further Philosophical Remarks about Semantics in Poland

The development of semantics in Poland was influenced by philosophers as well as mathematicians, that is, mathematical logicians. That Twardowski played a role was already mentioned above. On this, Tarski was explicit:

> Almost all researchers who pursue the philosophy of exact sciences in Poland are indirectly or directly disciples of Twardowski, although his own work could hardly be counted within this domain.
>
> (Tarski 1992, 20)

"Almost" occurs in this passage not simply as a matter a general caution. Chwistek here was one of the rare exception; he never studied with Twardowski. Now, there is an important difference between Chwistek's approach to semantics and that of Twardowski's students. Briefly, whereas Chwistek reduced semantics to a formal theory of expressions, logicians influenced by Twardowski looked upon referential relations as inherently involved in the use of language. In general, Polish philosophers, influenced by Brentano via Twardowski, transformed the intentional conception of mental acts into a referential treatment of language. This position has its roots in the conception we mentioned earlier, according to which intentionality is always involved in the use of linguistic expressions. Hence, language was inevitably viewed in Poland as a semantic matter and once again we note the crucial importance of Twardowski.

Polish logicians and philosophers were also influenced by Husserl, although whether this influence has much to do with what became known as "phenomenology" depends much on the meaning we associate to the term. I already mentioned that Leśniewski's idea of syntactic categories was modelled on Husserl notion of *Bedeutungkategorie*. It is also well known that antipsychologism in Poland was prompted by Husserl's criticism in the *Prolegomena to Pure Logic* (see Woleński 2003 for details).

When it comes to understanding Polish semantics we may however appeal to a more general framework. Jaakko Hintikka (see Hintikka 1988) and Kusch (see Kusch 1989) attempted to explain some historical facts and, in particular, the reason why some influential philosophers failed to develop a semantics by appealing to a distinction between language as calculus (LCA) and language as universal medium (LUM). The difference can be illustrated schematically in the following manner (see Kusch 1989, 6–7; see also Woleński 1997):

	LUM	*LCA*
(a)	semantics is not accessible;	semantics is accessible;
(b)	different systems of semantic relations are inconceivable;	different systems of semantic relations are conceivable;
(c)	model theory is rejected;	model theory is accepted;
(d)	semantic Kantianism is adopted;	semantic Kantianism is rejected;
(e)	metalanguage is illegitimate	metalanguage is legitimate;
(f)	truth as correspondence is not intelligible;	truth as correspondence is intelligible;
(g)	formalism is linked with the thesis: (*) semantics is not accessible;	formalism is linked with the thesis: semantics is accessible.

For Kusch and Hintikka, who defend LCA, language is a re-interpretable set of expressions; hence, language is not conceived, on this view, as a purely formal and thereby uninterpreted calculus. In general, according to LUM, semantics is simply ineffable, as stated in (*). As we immediately see from the comparison of the two accounts, they disagree on every point.

Hintikka and Kusch use the LUM/LCA distinction to paint a picture of contemporary philosophy of language: Frege, Russell and Wittgenstein represent the LUM account, Husserl, Gödel and Tarski the LCA view. In particular, they believe that LUM blocked the development of semantics because it forced philosopher to look at language as something that excludes any way of speaking about itself. This way of looking at the history of contemporary semantics is certainly illuminating, but it is also liable to some objections. In particular I think that when it comes to LCA, we should distinguish the global from the local approach. The local approach to LCA need not to be inconsistent with LUM. For example, one can consider fragments of language to be calculi while considering language as a whole to be a universal medium. If we agree to this qualification, then it seems that Frege was a "localist" in some cases, and a "universalist" as far as the integrity of language is concerned. I also have doubts as to whether Gödel accepted LCA in all details, because he was looking for a universal system of set theory, at least in the last years of his career. Thus, except for some hermeneuticians, such as Heidegger or Gadamer, only Wittgenstein (at least, in the *Tractatus*) remained a pure universalist. This is all the more paradoxical that while his idea that propositions represent facts contributed to semantics, perhaps against his own intentions, he rejected the view that one can meaningfully speak about relations holding between what is represented and what is representation. In any case, it seems that a discussion of LCA should always include the observation that language, on this view, is stratified into an infinite hierarchy of levels that can be re-interpretable at every stage. Tarski took this step, although, in his later works (see above), he simplified the picture by adopting only two levels.

Although the general philosophical climate in Poland was very favourable to semantics, it was not enough to sanction it entirely. In fact, many had reservations towards semantics in the philosophical world of the 1930s. Tarski described the situation in the following manner:

> Concepts from the domain of semantics have traditionally played a prominent part in the discussions of philosophers, logicians and philologists. Nevertheless they have long been regarded with a certain scepticism. From the historical point of view this scepticism is well founded; for, although the content of the semantical concepts, as they occur in colloquial language, is clear enough, yet all attempts to characterize this content more precisely have failed, and various discussions in which these concepts appeared and which were based on quite plausible and seemingly evident premises, have often lead to paradoxes and antinomies. It suffices to mention here the antinomy of the liar, the Grelling-Nelson antinomy of hererological terms and the Richard antinomy of definability.
>
> (Tarski 1936, 401)

Thus, for Tarski, the danger of antinomies was one cause for scepticism toward semantics. But other factors contributed to this sceptical attitude as well. Some members of the Vienna Circle (Neurath, for example) constantly accused semantics of introducing metaphysical elements (for a discussion of this and related matters, see Mancosu, this volume). This compelled semantics to show that it was methodologically legitimate (compare the title of Tarski 1936) and this was an intricate problem. Hilbert's metamathematics had produced solid devices to work with syntax in a finitary way. The latter, later enriched by Gödel's technique of arithmetization, prevented antinomies. In spite of compositionality and the connection between syntax and semantics, finitary methods were still insufficiently rich to convey semantic concepts. In this respect, the Polish mathematical heritage became very important for Tarski's "prosemantic" attitude. It had its origin in the Polish mathematical school with its focus on set theory, topology and their applications in other branches of mathematics. This school had no hostility towards infinitary methods as the following quotation clearly shows (Sierpiński had been holding this view since at least 1918):

> Still, apart from our personal inclination to accept the axiom of choice, we must take into consideration, in any case, its role in the Set Theory and in the Calculus. On the other hand, since the axiom of choice has been questioned by some mathematicians, it is important to know which theorems are proved with its aid and to realize the exact point at which the proof has been based on the axiom of choice; for it has frequently happened that various authors have made use of the axiom of choice in their proofs without being aware of it. And after all, even if no one questioned the axiom of choice, it would not be without interest to investigate which proofs are based on it and which theorems can be proved without its aid – this, as we know, is also done with regard to other axioms.
>
> (Sierpiński 1965, 95)

Thence the rule: use infinitary methods until a contradiction stemming from their use is proved. As far as metamathematics is concerned, Tarski described the situation in the following statement:

> As an essential contribution of the Polish school to the development of metamathematics one can regard the fact that from the very beginning it admitted into metamathematical research all fruitful methods, whether finitary or not.
>
> (Tarski 1954, 713)

And this is the point where semantics and metamathematics finally meet for the first time, so to speak, "on the record". It is, of course, not true that the notion of truth did not belong to the mathematical jargon before Tarski. In (Hilbert and Ackermann 1928) or (Carnap 1929), the concept was used informally, although essentially, for example in the formulation of the completeness problem. However, as Gödel remarked, and although he employed it himself in the formulation of the completeness theorem – and in the intuitive formulation of the first incompleteness theorem; more on this in what follows – it had not as yet been formally elaborated (see Gödel 1929, Gödel 1930). Only Tarski consciously combined philosophical and mathematical interests in insisting on elaborating semantic concepts with exactness. It was by no means accidental that the rudiments of the formal analysis of truth – the concepts of satisfaction and definability – first appeared in a paper in which some problems of descriptive topology, a domain where the transfinite plays an important role, were studied (see Tarski 1931).

When it comes to the use of infinitary methods, it is interesting to compare Tarski and Gödel (see also Woleński 2005). The latter in his memoirs insists that the general philosophical context of the 1920s and 1930s largely prevented the serious treatment of the concept of truth:

(a) I may add that my objectivist conception of mathematics and metamathematics in general, and of transfinite reasoning in particular, was fundamental also to my other work in logic. [...]. [...] it should be noted that the heuristic principle of my construction of undecidable number theoretic propositions in the formal systems of mathematics is the highly transfinite concept of 'objective' mathematical truth as *opposed* to that of 'demonstrability' [...] which was generally confused before my own and Tarski's work.

(Gödel 2003a, 398)

(b) I am still *perfectly convinced* that reluctance to use non-finitary concepts and arguments in metamathematics was the primary reason why the completeness proof was not given by Skolem or anybody else before my work.

(Gödel 2003a, 403–404)

(c) [...] in consequence of the philosophical prejudices of our times [...] a concept of mathematical truth as opposed to demonstrability was viewed with greatest suspicion and widely rejected as meaningless.

(Gödel 2003, 10)

Gödel's remarks about the putative prejudices (surprisingly enough, he struck out the passage in the draft of the letter to Balas) concerning non-finitary reasoning concerned logicians like Skolem, Hilbert or Herbrand, that is, leading figures in metamathematics in the 1920s and early 1930s. This attitude is clearly echoed in Gödel's paper on the incompleteness of arithmetic (see Gödel 1931) in which he gives the following informal explanation of the first incompleteness theorem, that is, the statement that if arithmetic is consistent, it is incomplete. Assume that arithmetic is consistent and consider the sentence (A): "(A) is unprovable". If it is true, then (A) is unprovable. If we further assume that (A) is false, then (A) is provable. Now, if we additionally assume that logic does not prove false sentences, that is, if we assume that logic is correct (sound), the negation of (A) is also not provable. Thus, arithmetic is incomplete, because there is a sentence such that, neither it, nor its negation, are provable. Gödel then adds:

The method of proof [of incompleteness – J. W] just explained can clearly be applied to any formal system which, first, when interpreted as representing a system of notions and propositions, has at its disposal sufficient means of expression to define the notions occurring in the argument above (in particular, the notion "provable formula") and in which, second, every provable formula is true in the interpretation considered. The purpose of carrying out the above proof with full precision in what follows is, among other things, to replace the second of the assumptions, just mentioned by a purely formal and much weaker one.

(Gödel 1931, 151)

Now, in the light of what precedes, what Gödel means with the idea that his "heuristic principle" relates to the notion of objective truth is clear. The exact proof of his theorem consists in showing that arithmetic has the means to represent metamathematical concepts, in particular, the notion of provability. The technique of arithmetrization enables one to formalize metamathematics inside formalized arithmetic. Thus, (A) is expressed in the language of arithmetic by purely syntactic devices. In the informal setting, there is little of consequence in saying: "if arithmetic is consistent, it is incomplete", because the informal proof refers to any system plus (A), and the consistency-assumption expresses only the trivial fact that not every sentence is provable. The matter changes when arithmetized metamathematics is taken into account. Now, the incompleteness theorem takes the form: if *S* is arithmetizable (contains arithmetic) and ω-consistent, it is incomplete. Once (A) is arithmetized, the assumption of consistency is enough to prove its unprovability. But in order to show the same about the negation of (A), a stronger assumption, namely that of *ω*-consistency must be made (I set aside the Rosser result that consistency suffices if (A) is modified). Thus, while the application of Gödel's techniques allows one to replace the semantic assumption of (A)'s truth with the syntactic assumption of the consistency of *S*, in order to replace the corresponding assumption for the negation of (A), one needs ω-consistency. Both assumptions are purely syntactic and, in consequence, the proof is also purely syntactic or, in other terms, proceeds by combinatorial methods. Now, a deep difference between Gödel's incompleteness theorem and Tarski's undefinability theorem is that while the former can be proved by mere finitary methods, the latter cannot. This shows that the difference between syntax and semantics is actually very deep: while the former is entirely finitary, the latter is not. Tarski, because he matured in a mathematical atmosphere fully tolerant of infinitary methods, had not the prejudices mentioned by Gödel and he could therefore make the decisive step toward semantics as a formal theory.

3.7 Final Remarks

The philosophical and mathematical triumph of semantics today is enormous. At least two important philosophers, Kazimierz Ajdukiewicz and Rudolf Carnap, changed their fundamental views under the influence of Tarski. Nowadays, the exposition of logic is much more frequently semantic than syntactic. Semantic ideas are used in philosophy of science, philosophy of language, ontology, epistemology, ethics and aesthetics. The entire realm of philosophy is penetrated by semantics. What's more, contemporary linguistics is strongly influenced by formal semantics. This, of course, does not mean that philosophy or linguistics are nothing beyond semantics. But it is difficult to imagine what would be the present shape of analytic philosophy without the development of semantics in the 1930s. The intellectual environment created by Polish philosophers and mathematicians, in this respect, was decisive.

References

Ajdukiewicz, K. (1930), 'Elementy teorii poznania, logiki formalnej i metodologii nauk T. Kotarbińskiego' [Review of Kotarbiński 1929], *Przegląd Filozoficzny* 23/1–2, 140–160; Eng. trans. in Kotarbiński 1929, 514–536.

Ajdukiewicz, K. (1931), 'O znaczeniu wyrażeń' (On the Meaning of Expressions), in *Księga Pamiątkowa Polskiego Towarzystwa Filozoficznego we Lwowie* (The Memorial Book of the

Polish Philosophical Society in Lvov), Polskie Towarzystwo Filozoficzne, Lwów; Eng. trans. (by J. Giedymin) in Ajdukiewicz 1978, 1–34.
Ajdukiewicz, K. (1936), 'Die syntaktische Konnexität', *Studia Philosophica* 1, 1–27; Eng. trans. (by H. Weber), in Ajdukiewicz 1978, 118–139.
Ajdukiewicz, K. (1978), *The Scientific World-Perspective and Other Essays*, 1930-1963, Dordrecht, D. Reidel.
Ajdukiewicz, K. (1993), 'Kategorie syntaktyczne i antynomie logiczne. Wykłady z semantyki logicznej w r. a. 1930/1931' (Syntactic Antinomies and Logical Antinomies. Lectures on Logical Semantics in the Academic Year 1930/1931), *Filozofia Nauki* 1/1, 163–182.
Betti, A (2004), 'Leśniewski's Early Liar, Tarski and Natural Language', *Annals of Pure and Applied Logic*, 127, 267–287.
Carnap R. (1929), *Abriss der Logisitik mit besonderer Berücksichtigung der Relationstheorie und ihrer Anwendungen*, Julius Springer, Wien.
Chwistek L. (1935), *Granice nauki. Zarys logiki i metodologii nauk ścisłych* (The Limits of Science. Outline of Logic and Methodology of the Exact Sciences), Lvov, Książnica-Atlas; Eng. trans. (by E. Brodie and A. Coleman), London, Kegan Paul, 1948.
Czeżowski, T. (1918), 'Imiona i zdania. Dwa odczyty' (Names and Sentences. Two Lectures), *Przegląd Filozoficzny* 21, 3–22.
Gödel, K. (1929), *Über die Vollständigkeit des Logikkalküls*, Wien Universität, repr. with Eng. trans. (by S. Bauer-Mengelberg) in Gödel 1986, 60–101.
Gödel, K. (1930), 'Die Vollständigkeit der Axiome des logischen Funktionenkalkül', *Monatshefte für Mathematik und Physik* 37, 349–360: repr. with Eng. trans. (by S. Bauer-Mengelberg) in Gödel 1986, 102–123.
Gödel, K. (1931), 'Über formal unentscheidbare Sätze der *Principia Mathematica* und verwandter Systeme I', *Monatshefte für Mathematik und Physik* 38, 173–198; repr. with Eng. trans. in Gödel 1986, 145–195.
Gödel, K. (1986), *Collected Works Volume I, Publications 1929-1936*, ed. S. Feferman et al., Oxford, Oxford University Press.
Gödel, K. (2003), *Collected Works Volume IV, Correspondence A-G*, ed. by S. Feferman et al., Oxford, Oxford University Press.
Gödel, K. (2003a), *Collected Works Volume IV, Correspondence A-G*, ed. by S. Feferman et al., Oxford, Oxford University Press.
Hilbert, D. and Ackermann, W., 1928, *Grundzüge der theoretischen Logik*, Springer, Berlin.
Hintikka, J. (1988), 'On the Development of Model-Theoretic Viewpoint in Logical Theory', *Synthese* 77, 1–36; repr. in Hintikka 1997, 104–139.
Hintikka, J. (1997), *Lingua Universalis vs. Calculus Ratiocinator. An Ultimate Presupposition of Twentieth-Century Philosophy* (J. Hintikka, Selected Papers 2), Dordrecht: Kluwer Academic Publishers.
Hintikka, J. (1997), *Lingua Universalis vs. Calculus Ratiocinator. An Ultimate Presupposition of Twentieth-Century Philosophy* (J. Hintikka, Selected Papers 2), Dordrecht: Kluwer Academic Publishers.
Kotarbiński, T. (1929), *Elementy teorii poznania, logiki formalnej i metodologii nauk* (Elements of Theory of Knowledge, Formal Logic and Methodology of Science), Lvov, Zakład Narodowy im. Ossolińskich; Eng. trans. (by O. Wojtasiewicz) as *Gnosiology. The Scientific Approach to the Theory of Knowledge*, Pergamon, Oxford, 1966.
Kusch, M. (1989), *Language as Calculus vs. Language as Universal Medium. A Study in Husserl, Heidegger and Gadamer*, Dordrecht, Kluwer.
Leśniewski S. (1913), 'Krytyka zasady wyłączonego środka' (The Critique of the Logical Principle of Excluded Middle), *Przegląd Filozoficzny* 16, 315–352; Eng. trans. in Leśniewski 1992, 47–75.
Leśniewski, S. (1929), 'Grundzüge eines neuen Systems der Grundlagen der Mathematik', *Fundamenta Mathematicae* 14, 1–81; Eng. trans. in Leśniewski 1992, 410–605.
Leśniewski, S. (1992), *Collected Works*, ed. and trans. by S. J. Surma, J. Strzednicki, and D. I. Barnett, Dordrecht, Kluwer.

Łukasiewicz J. (1915), 'O nauce' (On Science), in *Poradnik dla Samouków* (A Guide for Auto-Didacts) vol. 1, by S. Michalski (ed.), Warsaw, Heflich i Michalski, xv–xxxix.

Łukasiewicz J. (1920), 'Logika dwuwartościowa' (Two-Valued Logic), *Przegląd Filozoficzny* 23, 189–205; Eng. tr. (by O. Wojtasiewicz) in Łukasiewicz 1970, 89–109.

Łukasiewicz, J. (1970), *Selected Works*, L. Borkowski (ed.), North-Holland, Amsterdam.

Niedźwiedzka-Ossowska, M. (1925), 'Semantyka prof. St. Szobera' (Semantics of Prof. St. Szober), *Przegląd Filozoficzny* 28, 93–107.

Ossowski, S. 1926, 'Analiza pojęcia znaku' (An Analysis of the Concept of Sign), *Przegląd Filozoficzny* 29, 29–56); Eng. trans. (by O. Wojtasiewicz) in *Semantics in Poland 1894–1969*, J. Pelc (ed.), Dordrecht, D. Reidel, 1979, 164–177.

Siebel, M. (1996), *Der Begriff der Ableitbarkeit bei Bolzano*, Academia Verlag, Sankt Augustin.

Sierpiński, W. (1965), *Cardinal and Ordinal Numbers*, Państwowe Wydawnictwo Naukowe, Warsaw.

Szober, S. (1923), *Gramatyka języka polskiego* (Grammar of Polish Language), Lvov-Warsaw, Książnica.

Szober, S. (1924), *Zarys językoznawstwa ogólnego* (An Outline of General Linguistics), Warsaw, Wydawnictwo Towarzystwa Języka Polskiego.

Tarski, A. (1930–1931), 'O pojęciu prawdy w odniesieniu do sformalizowanych teorii dedukcyjnych' (On the Concept of Truth in Formalized Languages of Deductive Sciences), *Ruch Filozoficzny* 12, 210–211; repr. in Tarski 1986, vol. 4, 555–559.

Tarski, A. (1931), 'Sur les ensembles définissables de nombres réels I', *Fundamenta Mathematica* 17, 210–239; Eng. trans. (by J. H. Woodger) in Tarski 1983, 110–142.

Tarski, A. (1932), Der Wahrheitsbegriff in den Sprachen der deduktiven Disziplinen', *Akademie der Wissenschaften in Wien, Mathematisch-naturwissenschafliche Klasse, Akademischer Anzeiger* 69, 9–12; repr. in Tarski 1986, vol. 1, 613–617.

Tarski, A. (1933), *Pojęcie prawdy w językach nauk dedukcyjnych* (The Concept of Truth in Languages of Deductive Sciences), Warsaw, Towarzystwo Naukowe Warszawskie.

Tarski, A. (1936), 'Der Wahrheitsbegriff in den formalisierten Sprachen', *Studia Philosophica* 1, 152–278; repr. in Tarski 1986, vol. 2, 51–198; Eng. trans. (by J. H. Woodger) in Tarski 1983, 152–278.

Tarski, A. (1936a), 'Grundlegung der wissenschaftlichen Semantik', *Actes du Congrès International du Philosophie Scientifique*, vol. 3, Hermann, Paris 1938, 1–8; Eng. trans. (by J. H. Woodger) in Tarski 1956, 401–408.

Tarski, A. (1939), 'On Undecidable Sentences in Enlarged Systems of Logic and the Concept of Truth', *The Journal of Symbolic Logic* 4/3, 105–112.

Tarski, A. (1944), 'The Semantics Definition of Truth and the Foundations of Semantics', *Philosophy and Phenomenological Research* 4, 341–375.

Tarski, A. (1954), '[Contribution to the Discussion of P. Bernays, *Zur Beurteilung der Situation in der beweistheoretischen Forschung*]', *Revue Internationale de Philosophie* 8, 16–20; repr. in Tarski 1986, vol. 4, 713–714.

Tarski, A. (1962), 'Some Problems and Results Relevant to the Foundations of Set Theory', in *Logic, Methodology and the Philosophy of Science. Proceedings of the 1960 International Congress*, E. Nagel, P. Suppes, and A. Tarski (eds.), Stanford, Stanford University Press.

Tarski, A. (1983), *Logic, Semantics, Mathematics. Papers from 1923 to 1938*, 2nd edition, J. Corcoran (ed.), Indianapolis, Hackett Publishing Company.

Tarski, A. (1986), *Collected Papers*, vol. 1–4, S. R. Givant and R. N. McKenzie (eds.), Basel, Birkhäuser.

Twardowski K. (1894), *Zur Lehre vom Inhalt und Gegestand der Vorstellungen*, Hölder, Wien; Eng. trans. (by R. Grossmann), *On Content and Object of Presentations*, The Hague, Martinus Nijhoff, 1976.

Twardowski, K. (1912), 'O czynnościach i wytworach. Kilka uwag z pogranicza psychologii, gramatyki i logiki' (On Actions and Products. Some Remarks from the Boderline of Psychology, Grammar and Logic), in *Księga Pamiątkowa ku uczczeniu 250 rocznicy założenia*

Uniwersytetu Lwowskiego przez króla Jana Kazimierza (The Memorial Book for Celebration the 250th Anniversary of Establishing the Lvov University by the King John Casimir II), Lvov, Uniwersytet Lwowski; Eng. trans. (by A. Szylewicz) in Twardowski 1999, 103–132.
Twardowski, K. (1999), *On Actions, Products and Other Topics in Philosophy*, J. Brandl and J. Woleński (eds), Amsterdam, Rodopi.
Woleński, J. (1997), 'Husserl and the Development of Formal Semantics', *Philosophia Scientiae* 1/4, 151–158; repr. in Woleński 1999a, 110–114.
Woleński, J. (1999), 'Semantic Revolution: Rudolf Carnap, Kurt Gödel, Alfred Tarski', in *Alfred Tarski and the Vienna Circle*, J. Woleński and E. Köhler (eds.), Dordrecht, Kluwer, 1–15.
Woleński, J. (1999a), *Essays in the History of Logic and Logical Philosophy*, Jagiellonian University Press, Cracow.
Woleński, J. (2002), 'From Intentionality to Formal Semantics' (From Twardowski to Tarski), *Erkenntnis* 56(2002), 9–27.
Woleński, J. (2003), 'Psychologism and Metalogic', *Synthese* 137, 179–193.
Woleński, J. (2004), 'The Reception of Frege in Poland', *History and Philosophy of Logic* 25, 37–51; repr. in *Gottlob Frege: Critical Assessments of Leading Philosophers, vol. 1*, M. Beaney and E. Reck (eds.), London, Routledge, 290–309.
Woleński, J. (2005), 'Gödel, Tarski and Truth', *Revue Internationale de Philosophie* 59, 459–490.

Chapter 4
French and Polish Conventionalism

Anna Jedynak

French conventionalism originated at the turn of the 19th century. The views of Duhem, Poincaré and Le Roy are usually ranked under this heading and they will be presented in some detail in what follows. The purpose of this paper is to show the significance and considerable influence of French conventionalism on the development of philosophy and to present, in particular, its repercussions in Polish philosophy.

There is a number of different types of confusions and misinterpretations surrounding conventionalism as a whole. It is sometimes seen as a peculiar and rather insignificant episode in the history of philosophy of science. Its crucial role and marked impact on various philosophical ideas is sometimes ignored, especially its contribution to the anti-inductionistic and anti-positivistic turn. Critical opinions about it are quite common. Some see it as suggesting arbitrary solutions to problems whose resolution is warranted neither by analytical, nor by empirical knowledge. And since an arbitrary decision can easily lead to falsity, it would seem that conventionalism sanctions falsity.

These opinions are typically based on two misunderstandings and I would like to start by addressing them directly. The first mistake consists in treating conventionalism as a whole as if it concerned factual decisions, while we should in fact differentiate between two types of conventionalism: factual and linguistic conventionalism. Linguistic conventionalism ascribes the status of linguistic conventions to some sentences often otherwise regarded as based on experience. It is those conventions that make it possible to give an account of our experiences and to resolve factual problems. Linguistic conventionalism stresses the influence of linguistic conventions on the form of science and downplays the impact of empirical evidence. Factual conventionalism, on its part, is not interested in language, but in the research procedures and in the way scientists reach decisions when scientific statements in science are empirically underdetermined. It takes for granted that some procedures used for resolving certain factual problems which are empirically underdetermined are conventional.

A. Jedynak (✉)
Warsaw University, Warsaw, Poland
e-mail: annjed@wp.pl

S. Lapointe et al. (eds.), *The Golden Age of Polish Philosophy*, Logic, Epistemology, and the Unity of Science 16, DOI 10.1007/978-90-481-2401-5_5,

Drawing a clear distinction between linguistic and factual conventionalism is complicated by the fact that, although they deal with different issues, their areas of interest sometimes overlap. The same eminently acceptable scientific statements can be interpreted by the proponent of each type of conventionalism in a different way: in linguistic conventionalism, as a new meaning postulate meant to make previously vague terms more precise; in factual conventionalism, as an empirical statement expressed in an existing, unchanging language which may be justified – to some extent, but certainly not fully – by experience and accepted by virtue of a decision made by some scientists or scientific community.

Secondly, one could note that while the objection that its sanctions falsity may be justified as regards factual conventionalism, it does not apply to linguistic conventionalism. Misunderstandings would thus in any case arise from extending this objection to conventionalism as a whole, when their distinctiveness is insufficiently drawn. But the objection itself constitutes a misunderstanding: factual conventionalism alone does not sanction falsity, for it does not sanction anything at all. Conventionalism, including the factual one, is a descriptive conception and not a normative one, as some critics maintain. It describes the actual scientific procedure of accepting sentences, sometimes revealing their concealed character, it shows that scientific statements seemingly determined by experience are not necessarily so, and points to elements of decisions without which progress in science would be very limited. If scientific procedures reconstructed by factual conventionalism were to be doubted, we should either prove these procedures to be wrong (which would not be in accordance with contemporary tendencies in the philosophy of science), or strip science of procedures which are actually applied and useful. The latter would thus have consequences very similar to those of the neopositivistic criteria of meaning: if they were to be applied strictly, science would have to be significantly impoverished.

4.1 Pierre Duhem

It was Pierre Duhem who established factual conventionalism. He seldom uses the word "convention", and does not consider the status of sentences of various types, and neither does he resort to the notion of analyticity or similar associated ones. He considers (in the context of the history of physics) decisions made by scientists when theories are empirically underdetermined. The main points of his philosophy are as follows:

1. In the spirit of anti-verificationism, Duhem assumed that theories are not proved by experience, and that their acceptance is always a matter of decision. Nor are they proved, according to him, by the falsification of competing theories, since crucial experiments in physics are not possible for two reasons. Firstly, theoretical sentences are empirically tested not in isolation, but conjointly, in the context of various extra assumptions. Hence negative test results refute the whole conjunction of those sentences. It is the scientist who decides which one of them to reject and since she could each time have made a different decision, rejecting such and such elements of the tested whole is neither inevitable nor final. This

view, which express a form of methodological holism, is known as "Duhem's Thesis" and when generalized to all theories – not only physical ones – as the "Duhem-Quine Thesis". Secondly, even if experiments could univocally refute all but one of the competing theories considered by a scientist, this would still not provide an irrefutable justification for the non-refuted theory. For one can never be sure that the alternative of all the considered theories is true. On the contrary, it may be that none of them is true. Maybe the true one was never taken into account by the scientist to begin with.

2. The results of observations and experiment, as well as measurement procedures and devices are only understandable in the context of the whole theory. In that sense, theory precedes observation. Observations do not provide the basis for formulating the theory, but are merely the means to test it. Duhem's views are anti-inductionistic and, at the same time, antipsychologistic: following his approach, observational sentences are related not to impressions, but to things, their features and their relations.
3. Duhem distinguished practical facts, theoretical facts and theories. Practical facts are statements about measurement results, which are expressed in general observational terms and their meaning is formed spontaneously. Moreover, if these sentences themselves have a general character, they are called common-sense laws. One such law represents many practical facts. Theoretical facts are sentences formulated by the scientist in theoretical terms, and their meaning is formed intentionally and precisely. Whenever these sentences themselves have a general character, they are laws of *theoretical* physics. One such law represents many ordinary or common-sense laws. They result from idealization: the scientist has the right to choose theoretical facts on the basis of the same observations, in the narrow limits of measurement error. And finally, the theory consists of few principles that conjointly entail consequences translatable into statements concerning physical features of things, using methods for defining and measuring these features in the given theoretical context. Within the limits of measurement errors, these statements should be consistent with theoretical laws, and this provides an empirical test for the theory. The theory is constituted by many theoretical laws, which – while otherwise unconnected – are to be considered as its consequences. The same laws can be constitutive of alternative theories. The formulation of theoretical laws and of theories is governed by the principle of economy of thinking (in Mach's sense).
4. The cognitive value of laws and theories according to Duhem can be understood as follows. Common laws can be assessed as true or false. They are permanent and absolute, but not precise. Theoretical laws have cognitive value as approximate views of reality, but do not have a logical value in an absolute sense. Whether or not they have an approximate logical value according to Duhem is an open matter. If we can ascribe one to them, it is typically only as a result of the truth or the falsity of the common-sense laws entailed by them. The cognitive value of a theory according to Duhem can be interpreted both in the spirit of realism and of instrumentalism. He regarded the explanatory aspect of a theory as a metaphysical superstructure, logically independent of the other aspects. He believed that physics should limit its cognitive aspirations to the observable

surface of phenomena and, presumably in order to reconcile his attitude as a scientist with his attitude as a Catholic, he assumed that both metaphysical and religious doctrines say something about reality, but not scientific ones (this view is completely opposite to the neopositivistic one), so the latter cannot be inconsistent with the former. On this account, a theory has therefore no cognitive value, but only a practical one. These are the views of Duhem the instrumentalist. Duhem the realist claimed, that although the realistic position is not justified by the logical reconstruction of a theory, it is justified by the history of physics, which shows its heuristic and predictive role. A theory is not only a representation and classification of laws; it is their *natural* classification, for ideal relations perceived by the mind between abstract concepts correspond to real relations between entities. Hence, the logical order of laws should reflect the ontological order (which does not seem to be consistent with Duhem's conviction that alternative theories can represent the same laws).

Duhem's position with regard to the realism/instrumentalism controversy can be connected with the problem of the empirical testing of theories. For Duhem, the accuracy or inaccuracy of predictions derived from a theory determines (though not conclusively) whether the theory refers to reality or to a fiction. A theory that is inconsistent with experience is not false, but constitutes a physically uninterpreted language (the instrumentalistic theme) whereas a theory consistent with experience is not only physically meaningful, but also empirically confirmed (the realistic theme). Both instrumentalism and realism are acceptable. The choice between them depends in a given situation on the strength of the empirical evidence supporting the theory: when the support is strong realism is better, otherwise instrumentalism is better.

5. Apart from methodological holism we find in Duhem's thought the seeds of the kind of semantical holism later developed by Quine: the content of laws and any interpretation of experience can only be understood in the context of the whole theory. However, Duhem's conventionalism is not a linguistic one: in his view theories contain no sentences that are immune to empirical observations (i.e. analytic statements). He did not disclose any scientific sentences as purely linguistic conventions and did not bring this problem up at all. It is always the theory as a whole that is subject to empirical revision, even if it serves to define terms used for formulating the laws. (Quine went on developing this theme, using the famous metaphor of greyness he expressed his belief in the indistinguishableness of analytic and synthetic components in science). Duhem's conventionalism is a factual one.

4.2 Henri Poincaré

Poincaré's conventionalism has two seemingly divergent goals: to justify his belief concerning the cumulative development of science and to pave the way for relativistic physics, questioning the Newtonian mechanics supported by hundreds of years of practice as well as by Kant's authority. His views can be summarized in the following four points.

1. In order to undermine the Newtonian concepts of absolute space and time, Poincaré argued that they are empirically ungraspable. Geometry has no empirical interpretation and is not an empirical science. Its theorems have the status of more or less convenient conventions. The latter could have been different, as non-Euclidean geometries show. Our geometry is neither true nor false; it is merely convenient given our psychophysical constitution. Besides, the common sense understanding of the identity of time intervals is vague. For scientific purposes, it has been conventionally defined in such a way as to make the laws of physics that refer to the concept of time as simple as possible.
2. The extent of conventions in natural sciences is not as wide as in geometry. For natural sciences, according to Poincaré, retain their empirical character. Some – but not all – natural laws are not factual ones, but rather serve as definitions of important notions (e.g. Newtonian laws). They were originally meant as empirical laws, but the notions used to express them were drawn from colloquial language and although they might also have been defined operationally, they remained in any case vague. The better these laws were being confirmed empirically, the stronger was the need to make the notions involved more precise. And so, from a certain point onwards, empirical laws begin to serve as definitional principles whose purpose is to strictly elucidate the notions involved and are as such immune to empirical testing. In the face of experience they can be revoked as inconvenient, defining unnecessary or even empty notions, but not as false. Poincaré maintained that principles in physics are useful if there are few of them. Apart from them there are also other laws that preserve their empirical character. Some laws have a double status: with reference to ideal conditions they function as principles, and with reference to empirically graspable conditions they function as experiential laws.
3. Poincaré ascribed the status of definitions to the scientific components he considered to be conventions, but he did not elaborate on this issue, leaving open two possible interpretations. According to the first one attributed to Russell, what Poincaré has in mind is a new type of sentences that were left untouched by Kant. The conventions-definitions would be neither synthetic (for they have no logical value) nor analytic (for they can be established arbitrarily). According to the second interpretation, Poincaré would consider conventions-definitions to be analytic sentences. We can supplement the second interpretation by saying that conventions-definitions differ from other definitions in their being projective and not reports. Just as other projective definitions, they have no logical value in the initial language which is to be enriched by them, whereas they do have a logical value in the richer language that results from adding them to the initial language. Before it is sanctioned, a convention remains arbitrary, but after it has been sanctioned it holds at least until the language is again modified. It is then undoubtedly an analytic sentence, that is, a sentence true in virtue of the meanings it itself implicitly ascribes to certain expressions. (Poincaré himself

used the concept of truth ambiguously: in a narrow sense he only ascribed it to empirical sentences and, in a wider sense, to analytic sentences as well). In my view the latter interpretation, which is attributed to Popper, is not only right, but also fruitful. It allows one to acknowledge in Poincaré's thought the sources of the widely discussed problem of an adequate distinction of analytic and synthetic components in science, whereas the former (Russellian) interpretation has little interest since Russell does not investigate the new type of sentences and does not tell whether Kant was right or not to ignore them. As for mathematics, Poincaré did not consider it to be analytic, but synthetic a priori (for he ascribed this status to the principle of mathematical induction, which he considered fundamental to mathematics).

4. Poincaré's conventionalism is linguistic, not factual. He considered conventions to be not substitute for the justification of empirical sentences, but precondition of such justification. Conventions are needed to form a language in which certain problems finally become decidable (because of a modification of their former meaning). We use this language to question nature and interpret the results of appropriately arranged experiments. As a result, cognition is neither passive nor receptive. Empirical sciences are supposed to reveal the harmony of the world, namely the regularities that govern it. However, only relations between the contents of perception are accessible to scientific cognition and not the contents of perception as such, since the latter are subjective and incommunicable, and only the former are intersubjective and communicable. Thus, science is supposed to reveal relations between observable phenomena. Theories are often used for this purpose. Theoretical terms are conventionally defined through their connection with certain observational terms and they serve to express certain relations to other observational terms – and it is the resulting statements that acquire empirical content.

According to Poincaré, a scientific fact is simply an ordinary fact when referred to in the language of science containing theoretical terms. Duhem argued against this: he argued that a theoretical fact has many observational consequences, which are not identical. So one cannot assume that descriptions of those consequences have one theoretical translation in common. Moreover, the same observational phenomena can be explained within different theories.

Poincaré's holism is different from Duhem's. For one thing, Poincaré believed that some laws (though not all of them) could be decided through a crucial experiment. He postulated gradual acceptance of successive laws in order to avoid testing them conjointly. But, he nonetheless adhered to a generalization (different than the Quinean one) of Duhem's thesis: if, on the basis of accepted linguistic conventions (including metric ones), experience suggests rejection of certain laws, a scientist can either reject them or change the conventions (i.e., the language). According to Duhem, the scientist decides which empirical law to reject, while according to Poincaré, what she decides is whether to revoke the empirical law or the linguistic conventions. This difference is a result of the fact that Poincaré – by contrast to Duhem – differentiated between empirical laws and conventions-definitions.

Poincaré maintained that one should neither indiscriminately abide by conventions, nor change them inconsiderately. If changes are needed, empirical laws are to be changed first. When laws and conventions are tested conjointly, then only the laws should be tested with respect to truth, while the conventions should be tested with regard to their usefulness.

4.3 Édouard Le Roy

Le Roy was a follower of Henri Bergson and he conceived of reality accordingly: constantly changing, dynamic, alive and unique. Neither scientific knowledge, nor common sense knowledge can grasp it. Constancy in science is only approximate – a constant is really a variable with a narrow field of variability. This theme was present also in Duhem and Poincaré, however Le Roy drew much more radical consequences from it. He claimed that:

1. All scientific laws are (within the limits of measurement errors) arbitrary and accidental. They presuppose a certain constant order and schematization, while reality submits neither to constant order, nor to schematization. Le Roy believed that science cannot discover necessary connections: it can only decree them. (This view undermines the Kantian idea of knowledge that is necessary, universal and also real). Scientific laws express certain relations, and these relations exist only from the mind's perspective. Laws have no cognitive value.
2. Both scientific and common sense cognition are inadequate when it comes to reality. Common sense "creates" facts through the segmentation (in one of many possible ways) of direct data and by isolating individuals from them. Then common sense classifies things on the basis of certain similarities (also chosen among many possible ones) between them, and this results in the creation of notions. The manner of segmentation of direct data is chosen on the basis of our utilitarian needs. Both perception and language are adjusted to the best possible fulfilment of those needs. The same happens in science. On the basis of the facts decreed by common sense, the scientist creates scientific facts, in one of many possible manners, and also on the basis of their practical usefulness for science. We can differentiate two types of laws in science: laws as definitions, general and strict but unverifiable, and laws as practical rules that govern our actions and should be assessed not in terms of truth but in terms of efficiency.
3. The success of common sense cognition and scientific cognition does not establish their cognitive value. Science boasts its successes but hides its defeats. Success results from the fact that science is developed just to fulfil our utilitarian needs, so it proves useful in the area for which it was created. It enables us to act efficiently (laws as rules) or to create a vision of reality (laws as definitions), but it does not enable us to grasp this reality just as it is. These views of Le Roy place him closer to instrumentalism than Duhem and Poincaré. He believed that what we can ultimately know is based on the laws which merely express certain tendencies in things and that reality is such that it is amenable to many proposed

descriptions. However, such a description only touches the surface of reality and presents a deformed view thereof, not allowing us to reach its core. Certain other formulations are equally valid but not as useful. On this point, Le Roy met with a firm rebuttal from Duhem and Poincaré, who spoke up for the empirical character and cognitive value of natural sciences. In spite of their intentions however, the edge of their polemics is weakened by their own frequent utterances in the spirit of instrumentalism, so the differences between them and Le Roy do not seem to be that big.

4. Experience depends on laws, not the other way around, because it is the laws that shape experience. Theory precedes observation. Theory enables us to formulate questions we would like to ask about nature, to construct measuring instruments and interpret their functioning and results. Particular laws and theories are interconnected and, frequently, referring to one of them means referring to the whole body of our knowledge. Experience enables us to "touch" reality only in the context of theory. On the other hand, experience created on the ground of a theory is referred to for the purpose of confirming that theory. Thus science is based on vicious circles, and that should strengthen our conviction about its conventional character.
5. Neither common sense nor science can provide knowledge of reality in its actual essence. However, it is possible to acquire such knowledge using "intuition in metaphysical cognition". The latter was a favoured topic for Le Roy. He was involved in the philosophy of science only in order to study the mental contribution to the act of cognition governed by utilitarian needs in order to be in a position to deny its influence when turning to metaphysical cognition.

In spite of some polemics, the three authors described above have a lot in common – the most heated disputes about specific details often occur between those whose positions differ relatively little. All three questioned rudimentary commonsense views on the nature of empirical knowledge. They believed that the mind's contribution to scientific cognition is greater and the factual input smaller than it was previously admitted, that observations are shaped by theoretical assumptions and that the inaccuracy of metric procedures is an important consideration for the philosophy of science. They all distinguished ordinary from scientific facts and all of them are close to instrumentalism and holism.

The role of French conventionalism and its influence is very unjustly downplayed in the philosophy of science. In what follows, I wish to point to some ten points concerning the influence of conventionalism on the discipline. For one thing, conventionalism was much less widespread than neopositivism which developed a generation later. Nonetheless, it was subject to criticism, for its postulates were too radical and utopian. Since the anti-positivistic and anti-inductionistic turn, conventionalism has grown gradually weaker. Although it is seldom mentioned, the most important aspects of this criticism were already present in French factual conventionalism. While stressing the empirical underdetermination of scientific laws, it promotes anti-verificationism (experience does not univocally prove scientific laws) and holistic anti-falsificationism (experience does not univocally refute scientific laws). It also anticipates the idea of theory-ladeness of experience and thus, the lack of a stable, empirical foundation for cognition and puts into question the obviousness of the explanatory role of scientific theories.

Linguistic conventionalism also sets the stage for an adequate distinction between the synthetic and analytic components in scientific laws. Poincaré and Le Roy questioned the view that all natural knowledge is factual and fully established by experience. However, the views they formulated were exaggerated, sometimes demarcating too large a range of conventionalist elements (especially Le Roy). They did not acknowledge factual components within sentences they considered to be pure definitions. This problem was dealt with later, using the means of modern logic. Duhem's treatment of analyticity, on the other hand, faces another problem, for he ignored the difference between linguistic and factual components.

Linguistic conventionalism shares many features with analytical philosophy. Its attitude towards science was analogous to the attitude of analytical philosophy towards traditional philosophy as a whole: it revealed language problems in disciplines which seemed merely factual ones, and indicated that linguistic means were often required for resolving important questions. Conventionalism furthermore questioned the idea of a passive or receptive character of cognition, and stressed the active role of the mind, manifested for instance in linguistic conventions. Conventionalism thus anticipated the linguistic relativism that would be developed a few decades later. Linguistic relativism claims that the segmentation of direct data and the separation into individuals could be made in various, alternative ways, and that the latter result in various mutually untranslatable languages, based on different presuppositions and the creation of different visions of the world. Linguistic relativism is closely related to Kuhn's cognitive relativism. It is interesting that both cognitive relativism and its overcoming in the spirit of rationalism found their inspiration in different motives of conventionalism. Conventionalism claims that what changes in science are the conventions, the definitions of theoretical concepts, the measurement techniques, the form of laws and theories (the latter is a theme of interest to relativists). But, on the other hand, it is assumed that there is a certain "universal invariant", as Poincaré would put it, which is hidden under all those factors. This invariant concerns the constant relations between observable phenomena and is expressed in scientific laws (this is a theme of interest to rationalists). Nowadays, there is a certain convergence of rationalism and relativism. In the spirit of Poincaré, the thesis of the untranslatability of languages expressing alternative theories is being undermined on the grounds that those theories serve to describe the same events and to explain and predict the same phenomena.

One could argue that conventionalism contributed to some extent to the establishment of post-modern philosophy (incidentally, both originated in France). While conventionalism concerns only empirical and deductive sciences (sometimes including also common-sense cognition), postmodernism could be understood as an extension of conventionalism to the humanities in general. According to postmodernism, the humanities cannot be treated "impartially", that is, purely "receptively", but only from one's own cognitive perspective, which is what traditional conventionalism claimed with reference to the subject-matter of natural sciences. According to postmodernism, adopting such a cognitive perspective in the humanities, just like adopting suitable conventions in empirical sciences, does not deform cognition, but rather constitutes the necessary condition for dealing with its object. During the first half of the 20th century, philosophy of science and social sciences drifted further and further apart. The former sometimes raised doubts as to whether it could still be called philosophy, since it ignored traditional philosophical problems. The latter raised another kind of worries, namely whether it could be verbalised at all. Some painstakingly attempted to overcome this split between the two philosophical trends and Le Roy, who studied the problems of philosophy of science to pave the way for taking up existential problems, was one of them. Conventionalism is also not unconnected to chaos theory. Poincaré, for one, stressed the inaccuracy of measurements in empirical sciences, from which he derived many philosophical and methodological consequences. Some of those consequences (namely those claiming that the mind has to simplify and order empirical data according to some conventionally assumed rules) resulted in the development of conventionalism. And other consequences (those claiming that apparently identical causes could lead to enormous and unpredictable differences in effects) resulted in the seeds of chaos theory. What's more, factual conventionalism posed the question of the rationality of inductive procedures (especially statistical ones) and, more generally, of all problems in making decisions about accepting sentences. Finally, instrumentalism may be seen to have been initiated as an answer to the potential objection that decisions made in the spirit of factual conventionalism may sanction falsity. Thus conventionalism opened up the problem of the cognitive and explanative value of empirical theories.

In general, we can say that some of the most important discussions in the 20th century philosophy of science were already at play in French conventionalism, and like the latter, they tend to undermine epistemological fundamentalism. They concern relations between observations and theories, the differentiation between factual and linguistic components of empirical laws, the rationalism-relativism and realism-instrumentalism controversies, and the rationality of accepting sentences and of acting in situations of uncertainty. Although the participants of these discussions do not explicitly refer to conventionalism, the themes they deal with actually derive from it.

The situation was somewhat different in Poland where the historical impact of French conventionalism was multifaceted and can easily be documented. Kazimierz Ajdukiewicz (1890–1963), an eminent member of the Lvov-Warsaw School, called his epistemological position "radical conventionalism". His aim was to develop and express in a precise manner the philosophical intuitions he had adopted from Le Roy, namely that experience alone does not force us to accept any articulated statement and that we need, in addition to the latter, linguistic conventions which express the mind's a priori contribution to cognition. Languages representing different conceptual "viewpoints" of the same world are mutually untranslatable. (Ajdukiewicz considered Poincaré's views too moderate, since according to Poincaré conventions are only necessary in the case of scientific facts and not in the case of bare ones).

Ajdukiewicz begins by elaborating a conception of meaning, which is supposed to explicate a common notion of meaning and justify the view that there exist languages which, while they have completely different sets of meaning for their expressions, are nonetheless suitable for speaking about the same reality. He assumed that, like a deductive system, a natural language is governed by three types of rules. Axiomatic meaning-rules, on the one hand, require that a speaker accepts certain sentences unconditionally (e.g. "Every square has four angles"). Deductive meaning-rules, on the other hand, require that a speaker accept certain sentences in view of her previous acceptance of other sentences (e.g. the acceptance of the sentence "John is older than Mary" if the sentence "Mary is younger than John" was previously accepted). Finally, empirical meaning-rules require that a speaker accept certain sentences in view of certain empirical data (e.g. the acceptance of the sentence "It hurts" when one actually has a toothache).

Ajdukiewicz assumed that the meaning of an expression is its position amongst all the meaning-rules of a given language. This, in a way, is intuitive: since meaning-rules hold as a consequence of the meaning of expressions, then this meaning should be hidden amongst them and should manifest itself because of them. (Later A. Tarski would notice that this dependence holds only on one side: meaning univocally defines meaning-rules, but the meaning-rules do not univocally define meaning. Ajdukiewicz would, as a consequence, revise his view). So, according to the meaning-rule conception, expressions of the same language are synonymous if and only if their mutual exchange within all the meaning-rules leaves the latter unchanged (apart from their order). And expressions from different languages are synonymous if and only if the structures of all meaning-rules of both languages are isomorphic, and the expressions in question occupy – each of them in the respective meaning-rules of their languages – the same places. Hence, Ajdukiewicz adopted semantical holism: according to him, one cannot define the meaning of any isolated expression, taken apart from the whole set of meaning-rules, which also determine the meaning of all other expressions of the given language.

Ajdukiewicz took into account only closed and connected languages. A closed language is such that it cannot be enriched by any expressions with new meanings, that is, whose set of meaning-rules is complete. A connected language is such that it cannot be divided into isolated parts, because each two expressions are directly or indirectly meaning-related. Two expressions are directly meaning-related when they can be found within the same meaning-rule. They are meaning-related indirectly when one of them is the first element and the other the last element of a chain of expressions whose every two elements are directly meaning-related. Ajdukiewicz called the set of meanings in a closed and connected language a conceptual apparatus, and all its theses (that is, all the sentences which are to be accepted under the meaning-rules) – a conceptual world-picture. According to Ajdukiewicz, a world-picture depends on the choice of a conceptual apparatus. This statement expresses the thesis of radical conventionalism.

Radical conventionalism takes on however much originality when applied to closed and connected languages, and under the assumption that there exist at least two such languages with different conceptual apparata. Since a conceptual apparatus cannot be enriched by any new meaning, every attempt at changing it must lead to a radical change, that is, it must cause a shift towards a new, different apparatus. Two apparata are either identical or have no elements in common. None of them can be enriched without changing the meanings of old expressions, so it is impossible to form the connected sum of two languages related to those apparata.

As a consequence, the same reality can be fully described in different ways that cannot be mutually compared. There are different "conceptual viewpoints" providing alternative conceptual world-pictures, and the truth of one world-picture does not contradict the truth of another. None of them can be assessed from the viewpoint of the other, for each of them is conceptually ungraspable from the perspective of the other. Were we to compare a conceptual apparatus to glasses through which we may perceive the world, we could say, first, that we can only see something using glasses, and second, that we cannot look through more than one pair of glasses at a time.

Ajdukiewicz later gave up radical conventionalism – in part because of Tarski's above mentioned criticism, in part because he himself concluded that the concept of a closed and connected language was an empty one, and in part because of the rapid development of semantics. Indeed, radical conventionalism was meant to be free from semantics, for it was developed in an era during which logicians could not avoid semantic antinomies.

While Poincaré used his conventionalism to justify his views on the cumulative development of science, Ajdukiewicz saw radical conventionalism as a means to justify an opposite view on the rapid and stepwise development of science. According to radical conventionalism, the change of one thesis accepted under a meaning-rule within a scientific theory sometimes causes an avalanche of other meaning changes and leads to the creation of a completely new theory with a new language. (According to this view, this is what happened when one of the Euclidean axioms was changed or when the Einsteinean definition of simultaneity was adopted). According to Ajdukiewicz, this is a revolution in science, which brings not only the change of laws, but also the change of the notions needed to express laws. The conviction about the cumulative development of science often goes hand in hand with epistemological fundamentalism, while the view of its stepwise development tends to go together with relativism. A. Grobler, who is currently working on overcoming the absolutism-relativism opposition, has inspired lively discussions in Poland and provoked criticism (see e.g. (Woleński 1996) and a collection of papers in "Filozofia Nauki" 3–4 (2003): 49–161). The echoes of conventionalism in Grobler's philosophy are obvious. He claims that conceptual systems are created with respect to our cognitive needs, and the latter depend on our practical needs. Reality is open to an infinite number of conceptual orders, but no conceptual system can embrace all of them and something will always remain unexpressed because of the lack of suitable conceptual means.

Though Grobler is obviously influenced by Ajdukiewicz, he considers amendments to Ajdukiewicz's conception to be necessary for two reasons. First, Ajdukiewicz's position is based, as he himself came to acknowledge, on the empty notion of a closed and connected language. And second, Ajdukiewicz did not share the theory-ladeness thesis, quite common today. Since radical conventionalism was created when logical empiricism and the theory of sense data were widespread, Ajdukiewicz believed in a stable empirical foundation of cognition and later did not change his mind. Like the majority of philosophers, he ignored the theme of theory-ladeness of experience present in French conventionalism. (This however does not hold of the Lvovian Ludwig Fleck who, already in the thirties, had taken up the theoretical context of interpretation of observations, using the concept of "thought style" and who later influenced T. Kuhn. But Fleck may perhaps not be considered a fully-fledged Polish philosopher for he wrote mainly in German.) At any rate, Grobler investigates the problem of meaning with reference to open languages and, in particular, the problem whether the shift of meaning of expressions, when the latter are used in qualitatively new situations, is completely arbitrary or whether it is rather adjusted to their previous use and/or to communicative and cognitive needs. He defends the second option by developing an approach he names "fallibilistic essentialism", which is a (considerably more) moderate version of Ajdukiewicz's position that an open language can only be closed in one way.

As regards theory-ladeness, with which Ajdukiewicz did not deal, Grobler points out that the empirical data to which empirical meaning-rules refer have to be previously ordered, and that this is only possible on the basis of a preexisting conceptual system. Conceptual systems form a network of mutual relations and function within them locally, not globally. As our cognitive needs change, we change the system as well. However, this change does not need to be total, for conceptual systems are not (contrary to what Ajdukiewicz assumed) isolated from one another. After all, some conceptual systems are evaluated from the viewpoint of others, and various systems are applied in order to describe overlapping areas of experience. Grobler stresses the need to elaborate a theory of the development of conceptual systems. Indeed, radical conventionalism leaves open the problem of an adequate reconstruction of the transformation of a conceptual system, which takes place when one scientific theory is replaced by another.

In Ajdukiewicz's later philosophy, the empiricist attitude became increasingly sustained. He assumed that it was possible to base logic on experience if we weakened the meaning of logical constants so that the meaning-rules governing unconditional acceptance of logical laws as axioms did not hold any more. But how do we derive logical laws from experience? By contrast to what is the case with observational sentences, this cannot be done directly but only by testing logical laws conjointly with natural hypotheses. An accurate observational prediction derived from a natural hypothesis according to a logical law confirms both the law and the hypothesis, while the inaccuracy of the prediction falsifies the conjunction of both of them. It is up to the scientist which one of them to regard as false: only the law, only the hypothesis or both. The problem of making such a decision is quite similar to what's implied by Duhem's thesis about physical hypotheses being tested conjointly. And it is also similar to Quine's generalization of Duhem's thesis, and according to which natural hypotheses are to be tested conjointly with logic. However, there is a difference between Ajdukiewicz and the Duhem-Quine stance. Ajdukiewicz was not a methodological holist, because he did not assume the possibility of testing conjointly several purely natural hypotheses; he only admitted the possibility of testing (in a language, that did not yet exist but could be constructed) one such hypothesis conjointly with a law of logic.

Ajdukiewicz presented arguments to the effect that the justification of sentences commonly regarded as analytic has to refer to empirical premises. This is because such sentences could prove false if their existential consequences were shown to be false. For instance, the analytic sentence "The present king of France is a bachelor" is false, for according to a certain law of logic it entails a false consequence "there is a King of France at present". This argument corresponds to what French conventionalists meant by the idea that the contribution of linguistic and factual components to cognition is not as simple and straightforward as it might seem. They pointed to their mutual interdependence: conventions influence the approach to experience, and at the same time they are suggested by this very experience. At first, Ajdukiewicz emphasized only the former side of this dependence. Noticing the latter, however, eventually brought him closer to his French predecessors. Since his views came more than fifty years later than theirs, he could make use of a logical apparatus that was not available to them, carefully and meticulously resolving problems with the support of stronger arguments. The results he obtained initiated a lively discussion in Polish philosophy (a short outline of the discussion was given in (Przełęcki & Wójcicki 1968–9).

Drawing closer to radical empiricism, Ajdukiewicz also considered the possibility of constructing a language governed only by empirical meaning-rules, that is, a language in which no a priori knowledge could be expressed. The question was whether it would be possible to minimize consciously mental and linguistic influence on cognition, so that all knowledge would only be derived from experience. This is not unlike Le Roy's investigation of the mind's contribution to scientific cognition, which aimed at curtailing the latter in favour of metaphysical cognition.

Even though Ajdukiewicz did not refer to Duhem's views, he did take up in a very general way the main theme of factual conventionalism, namely the resolution of empirically underdetermined problems. In that context he did not use the word "conventionalism" and only gave a short and rough outline of the issue. The problem concerns the rationality of fallacious reasoning. Ajdukiewicz connected it with probability theory and claimed that fallacious reasoning is rational if the degree of acceptance of a conclusion is not bigger than the degree of reliability of the reasoning. His students (K. Szaniawski, H. Mortimer) dealt with the problem in detail, developing a logic of induction and decision theory and they made reference to Ajdukiewicz – as well as to Western scientists. They developed the notion that the acceptance of sentences in conditions of uncertainty always involves making decisions which should not be completely arbitrary, but are preceded by conventional acceptance of certain rational decision-making criteria.

Jerzy Giedymin, another student of Ajdukiewicz, also contributed to conventionalism in the philosophy of science. His investigations, in the 1960s, aimed at showing that conventional elements influence the acceptance of both theoretical and observational sentences. He was especially interested in problems related to theory-ladeness. He sought to prove that the acceptance of observational sentences depends not only on observations, but also on previous theoretical assumptions. A change in the latter influences the acceptance of observational sentences, even though the observations themselves do not change. Using pertinent examples drawn from the history of science, Giedymin documented retrospective rejections of observational sentences caused by change in theories that would happen even centuries later.

Other Polish philosophers also dealt with various topics related to conventionalism, but most of their theories were not directly influenced by French conventionalism. Poincaré's main philosophical books were translated into Polish at the beginning of 20th century, though they were never retranslated or published again. Le Roy and Duhem were never translated. Thus Poincaré was better known by Polish philosophers and influenced their views to a greater extent than other French conventionalists. For instance, Isidora Dąmbska shared Poincaré's conviction that the content of impressions cannot be communicated, so its cognition is not scientific. In general, Polish philosophers dealt with conventionalist problems as theoreticians rather than as historians. For instance, I. Dąmbska, J. Giedymin and A. Siemianowski wrote quite a lot on conventionalism, aiming at interpreting and understanding it properly, explicating conventionalistic terminology, defining the scope of conventions in various domains of human activities, and distinguishing its various versions and pointing out to their mutual relations. To some degree, their work helped to change, at least in Poland, the common view that conventionalism is a strange and dangerous philosophy which has no reservation against sanctioning falsity. In particular, I. Dąmbska and L. Chwistek compared – each of them independently – conventionalism and relativism. It is interesting that in Dąmbska's case the result was in favour of conventionalism, while Chwistek favoured relativism.

It was in Polish philosophy that the idea of metaconventionalism first appeared. Giedymin used this word in his preface to the English edition of Ajdukiewicz's papers (Ajdukiewicz 1978), but the seeds of the idea were already been present much earlier. In 1938 Dąmbska considered whether conventionalism could fall into regress when it sought to answer the question as to how to justify conventions and choose between them. She answered in the negative, arguing that conventionalism is a theory of language, not of metalanguage. And from there is only a step to ask whether anything should prevent conventionalism from extending to the metalevel. Giedymin would later assume, that the evolution of Ajdukiewicz's views from radical conventionalism to radical empiricism should be interpreted as a passage of conventionalism from the object level to the metalevel. Instead of asking about the role of the mind in cognition, at the metalevel, the question is whether it is possible to consciously control this role and, in particular, to minimize it. It is not only the meaning of expressions which depends on the choice of a language, but also one's view on what meaning is; not only the shape of the laws of logic, but also one's view on their status. Conventions decide not only which axioms to assume, but also whether to assume any at all. In short, at the metalevel, conventions influence more general and fundamental matters than at the object level.

The conception of metaconventionalism may lead to problems or challenges, and regress is one of them. On the other hand it may prove useful and inspire consideration for the way other philosophical views function at the metalevel. The metasubjectivist implications of ethical discourse constitute one issue that could profit from this approach. At the object level, one could argue, ethical opinions are subjective and justified by people's feelings. At the metalevel, however, each person may decide for themself what the status of their opinion is. (Jedynak 2003).

The seeds of metaconventionalism may already be seen in Duhem's work, in the fact that he did not univocally decide between realism and instrumentalism but tended to treat some theories as real and some other as fictional, depending how strong the empirical evidence supporting them was. According to this approach, it is not only the choice of the theory that is to some extent conventional, but also the choice of our position on its status.

References

Ajdukiewicz, Kazimierz (1978), *The Scientific World-Perspective and Other Essays, 1931–1963*, Jerzy Giedymin (ed.) Dordrecht/Boston, Reidel.

Ajdukiewicz, Kazimierz (1974), *Pragmatic Logic*, Dordrecht, Reidel.

Chwistek, Leon (1935), *Granice nauki* (*The Limits of Science*), Lvov; Eng. trans. *The Limits of Science. Outline of logic and of the methodology of the exact sciences*, London, Kegan Paul, Trench, Trubner, 1948.

Dąmbska, Izydora (1938), 'Konwencjonalizm a relatywizm', (Conventionalism versus relativism) *Kwartalnik Filozoficzny* 14–15 (1937–8), 328–37.

Dąmbska, Izydora (1975), *O konwencjach i konwencjonalizmie*, (On Conventions and on Conventionalism), Wroclaw/Warszawa, Ossolineum, 1975.

Duhem, Pierre (1914), *La théorie physique, son objet, sa structure,* Paris, Chevalier et Rivière.

Giedymin, Jerzy (1966), 'O teoretycznym sensie tzw. terminów i zdań obserwacyjnych' (On the Theoretic Sense of so called Observational Terms and Sentences), in *Teoria i doświadczenie* (*Theory and Experience*), Helena Eilstein and Marian Przełęcki (eds.), Warsaw, PWN, 1966, 91–109.

Giedymin, Jerzy (1978), 'Radical Conventionalism, its Background and Evolution: Poincaré, Le Roy, Ajdukiewicz'. Editor's Introduction to (Ajdukiewicz 1978), IX–LIII.

Grobler, Adam (2000), *Prawda a względność* (Truth versus Relativity), Cracow, Aureus.

Jedynak, Anna (2003), 'Pluralizm metaetyczny', (Metaethical Pluralism), *Etyka* 36, 77–86.

Le Roy, Édouard (1899a), 'Science et philosophie', *Revue de métaphysique et de morale* 7, 375–425.

Le Roy, Édouard (1899b), 'Science et philosophie', *Revue de métaphysique et de morale* 7, 503–562.

Le Roy, Édouard (1899c), 'Science et philosophie', *Revue de métaphysique et de morale* 7, 708–731.

Le Roy, Édouard (1899d), 'Science et philosophie', *Revue de métaphysique et de morale* 8, 1900, 37–72.

Le Roy, Édouard (1900) 'La science positive et les philosophies de la liberté'. in*Bibliothèque du Congrès international de Philosophie 1, Paris, 313–341.*

Le Roy, Édouard (1901a) 'Sur quelques objections adressées à la nouvelle philosophie', *Revue de métaphysique et de morale* 9, *407–432.*

Le Roy, Édouard (1901b), 'Un positivisme nouveau', *Revue de métaphysique et de morale 9, 138–153.*

Le Roy, Édouard (1901c), 'De la valeur objective des lois physiques', *Bulletin de la Société française de la philosophie*, Paris, séance du 28.03.1901.

Le Roy, Édouard (1912), *Une philosophie nouvelle. Henri Bergson*, Paris, Félix Alcan.

Poincaré, Henri (1902) *La Science et l'hypothèse*, Paris, Flammarion.

Poincaré, Henri (1905) *La valeur de la science*, Paris, Flammarion.

Poincaré, Henri (1908) *Science et méthode*, Paris, Flammarion.
Przełęcki, Marian & Ryszard Wójcicki (1968) 'The Problem of Analyticity', *Synthese* 19, (1968–9). 374–399.
Siemianowski, Andrzej (1989), *Zasady konwencjonalistycznej filozofii nauki*, Warszawa, PWN.
Woleński, Jan (1996), 'Nauka i prawda' (Science and Truth), *Znak* 9.

Part II
Philosophy of Logic and Mathematics

Chapter 5
A Philosophy of Many-Valued Logic. The Third Logical Value and Beyond

Grzegorz Malinowski

5.1 Introduction

The roots of many-valued logic lie in Aristotle's (4th century BC) discussion of *future contingents* and of tomorrow's famous sea battle. Similar concerns can be found in medieval philosophy, in Duns Scot, William of Ockham and Peter de Rivo (Louvain). However, one needs to wait until the turn of the twentieth century to see the first attempts at creating non-classical – mainly three-valued – systems. By the late 1890s, Hugh MacColl had presented his so-called "three-dimensional logic", Charles S. Peirce (1839–1914) was working on "trychotomic mathematics" and Nicolai A. Vasiliev was developing a system in which propositions can be "affirmative", "negative" or "indifferent".

The golden age of "many-valuedness" however began in the 1920s with the works of Jan Łukasiewicz (1920) and Emil Post (1920, 1921). Our discussion focuses on Łukasiewicz's non-classical logics. Special attention is paid to Łukasiewicz' three-valued propositional logic, the first system of the kind. We briefly go over Łukasiewicz motivation for introducing a third value next to truth and falsity. Next, we discuss some of the consequences of Łukasiewicz' construction and some of the limits of many-valued logic as a whole. The last part of the text deals with criticisms and recent interpretations of Łukasiewicz' third value and the general interest of many-valuedness in logic.

5.2 Łukasiewicz and the Lvov-Warsaw School

Jan Łukasiewicz is the co-originator and a prominent representative of the Lvov-Warsaw philosophical school. This is not a trivial remark since the discovery of many-valued logic is a direct result of some of the debates carried out within the School. In particular, among the factors that prompted Łukasiewicz' views on logics

G. Malinowski (✉)
Uniwersytet Łódzki, Łodz, Poland
e-mail: gregmal@uni.lodz.pl

S. Lapointe et al. (eds.), *The Golden Age of Polish Philosophy,* Logic, Epistemology, and the Unity of Science 16, DOI 10.1007/978-90-481-2401-5_6,

and which persuaded him to abandon the classical perspective, one may mention the following three:

1. The discussion, within the Lvov-Warsaw School, of the general theory of objects that had been proposed by Brentano, Twardowski and Meinong.
2. Łukasiewicz's investigation into the problems of induction and the theory of probability.
3. Łukasiewicz's examination of the question of determinism, indeterminism and related problems concerning causality and modality.

Ad (1). The general theory of objects coined towards the end of the 19th century by Meinong postulates the existence of contradictory objects, i.e. objects having contradictory properties, such as "round squares". Meinong insisted that we have to accept that contradictory objects exist and argued that assuming the non-existence of contradictory objects results in the inability to utter certain true propositions such as, for instance, propositions stating that non-existent objects are not objects. Łukasiewicz himself was not avert to the theory of contradictory objects – he not only advocated it but even leaned towards the belief that non-contradictory objects do not exist – and this became relevant to his treatment of antinomies. Russell's discovery of the paradox in set theory, involving the "set of all sets, which are not their own elements", $R = \{x : x \notin x\}$, was acknowledged by Łukasiewicz, who seems to have, at least initially, blamed the problem on the classical principle of contradiction (see Łukasiewicz 1910).

Ad (2). Łukasiewicz's investigation into the problem of induction and the theory of probability led him to take his distance from the "embarrassing" principle of contradiction (see Łukasiewicz 1913) and to classify as "indefinite" propositions containing free nominal variables. To such propositions Łukasiewicz' assigned fractional values, which were supposed to indicate the proportions between the number of actual values of a variable verifying a proposition and the number of all possible values of that variable. Under this conception, "logical values" are relative: they depend on the set of individuals actually evaluated. So, for example, the value of the proposition $x^2 = 1$ amounts to $^2/_3$ in the set $\{-1, 0, 1\}$ and to $^2/_5$ in the set $\{-2, -1, 0, 1, 2\}$. The idea, which he took over directly from Bolzano (1837, §147),[1] could have seemed attractive and is not entirely foreign to the probabilistic approach.[2] What is relevant here does not concern Łukasiewicz' motives for combining the problems of many-valuedness with that of probability. The crucial fact is that already in 1913 Łukasiewicz had employed the concept of a logical value in an unorthodox manner.

[1] In the same book, Łukasiewicz criticizes the notion of a variable Bolzano introduces in §147 of the *Wissenschaftslehre* in connection with his definition of "Gültigkeit" where he sets other foundation for a calculus of probability defined along the same peculiar lines as Łukasiewicz.

[2] Since infinite sets of individuals are not admitted, Łukasiewicz's suggestion cannot be taken seriously within any actual theory of probability. We owe some attempts to improve the idea to Zawirski (1934).

Ad (3). Łukasiewicz' studies previous to the construction of a three-valued logic touched upon determinism, indeterminism and some related problems concerning causality and modalities, i.e., possibility and necessity (see Łukasiewicz 1906, 1910). Some historians of logic suspect that Łukasiewicz was influenced by the debate concerning freedom and creativity. Notoriously, Kotarbiński (1913) saw the need to revise two-valued logic on the grounds that it seems to interfere with the freedom of human thought. Łukasiewicz, a fierce follower of indeterminism, introduced the third logical value to be assigned to non-determined propositions; specifically, to propositions describing causal future events, i.e., *future contingents* (see also the Introduction).

5.3 Three-Valued Logic

The earliest remarks about three-valued propositional calculus can be found in the "Farewell Lecture" given by Łukasiewicz on 7th of March, 1918 at the University of Warsaw. Next came the seminal paper "On Three-Valued Logic" (Łukasiewicz 1920), which briefly motivates the need for a new type of logical construction, casts off the principle of bivalence and provides an outline of three-valued logic. Łukasiewicz (1930) analyses the sentence "I shall be in Warsaw in a year" and states that, at the moment of it's utterance, its value (truth or falsity) is not settled. He suggests that sentences of this kind pertain to a "third" logical category. To the two classical values 0 and 1, he added an intermediate logical value $^1/_2$ interpreted as *possibility* or *indeterminacy* – the former of these options was subsequently repudiated by Łukasiewicz who had come to a different and deeper understanding of modalities. On the basis of this intuition, Łukasiewicz extends the classical interpretation of the negation ($\neg$) and implication ($\rightarrow$) operators, which results in the following tables[3]:

x	$-x$
0	1
$^1/_2$	$^1/_2$
1	0

$\rightarrow$	0	$^1/_2$	1
0	1	1	1
$^1/_2$	$^1/_2$	1	1
1	0	$^1/_2$	1

The other relevant connectives, disjunction ($\vee$), conjunction ($\wedge$) and equivalence ($\equiv$) are introduced by the following definitions:

$$\alpha \vee \beta = (\alpha \rightarrow \beta) \rightarrow \beta$$
$$\alpha \wedge \beta = \neg(\neg\alpha \vee \neg\beta)$$
$$\alpha \equiv \beta = (\alpha \rightarrow \beta) \wedge (\beta \rightarrow \alpha),$$

[3] The truth-tables of binary connectives are viewed as follows: the (logical) value of α the first vertical line, the value of β in the first horizontal line and the value of $\alpha^*\beta$ at the intersection of the two lines.

leading to the following truth tables:

$\vee$	0	$^1/_2$	1
0	0	$^1/_2$	1
$^1/_2$	$^1/_2$	$^1/_2$	1
1	1	1	1

$\wedge$	0	$^1/_2$	1
0	0	0	0
$^1/_2$	0	$^1/_2$	$^1/_2$
1	0	$^1/_2$	1

$\equiv$	0	$^1/_2$	1
0	1	$^1/_2$	0
$^1/_2$	$^1/_2$	1	$^1/_2$
1	0	$^1/_2$	1

A valuation v of the set of formulas *For* in Łukasiewicz three-valued logic is any function $v : For \rightarrow \{0, {}^1/_2, 1\}$ "compatible" with the above tables. Accordingly, a tautology in Łukasiewicz's three-valued logic is a formula which, under any admissible valuation v, takes on the *designated value* 1.

System Ł3 – "Ł3" being a designation for Łukasiewicz' system of a three-valued propositional calculus – differs radically from its classical counterparts. On the one hand, some important laws of classical logic are not tautologies in Ł3. On the other hand, some classically contradictory formulae[4] are consistent in Łukasiewicz's logic. Among the formulae of the first kind there is:

1. $p \vee \neg p$ (*law of the excluded middle*)
2. $\neg(p \wedge \neg p)$ (*principle of non-contradiction*).

To verify that these formulas are not tautological in Ł3, it suffices to see that any valuation v ascribing $^1/_2$ to p yields a value of $^1/_2$. On the other hand, we have the important formula of the second kind:

3. $p \equiv \neg p$,

which takes the value 1 for the valuations specified above.

The refutation of the law of the excluded middle and the principle of non-contradiction was intended to epitomize the principles of indeterminism since, as Łukasiewicz puts it, both the disjunction and the conjunction of two possible propositions are possible propositions (and nothing else). The consistency of (3) supports the claim that three-valued logic is adapted to the formalization of reasoning about contradictory objects and, as Łukasiewicz remarks, Russell's paradox of the set of all sets that are not their own elements ceases to be an antinomy in Ł3. True, the definition of the Russell's set R entails the paradox

$$Z \in Z \equiv Z \notin Z.$$

However, the latter formula is a substitution of $p \equiv \neg p$ and is thus consistent in Łukasiewicz logic.

5.4 The Third Value – Its Critiques and Recent Interpretations

As early as 1938, a serious blow was inflicted on Łukasiewicz's conception. Gonseth noticed (see Gonseth 1941) that the formal characterization of the connectives in Łukasiewicz logic is not compatible with the interpretation of the third logical value

[4] That is, formulae taking 0 for arbitrary valuation.

Łukasiewicz had suggested, that is, it can be interpreted neither as possibility nor as indetermination. Gonseth's argument is both sound and straightforward. Consider two propositions α and $\neg\alpha$. Whenever α is undetermined, so is $\neg\alpha$, and then, according to the table of conjunction, $\alpha \wedge \neg\alpha$ is undetermined, which contradicts the intuition since, independently from α's content, $\alpha \wedge \neg\alpha$ is false. The argument concerning Łukasiewicz's treatment of disjunction goes along the same lines, showing similar problems with the valuation of $\alpha \vee \neg\alpha$ which is not supposed to be a tautology in Ł3. According to Gonseth, Łukasiewicz's original interpretations result from his neglecting the mutual dependence of all undetermined, or possible, propositions. The example above shows that this assumption cannot be handled.

Urquhart (1986), to mention another well know critique, shows that the interpretation of Łukasiewicz's third logical value as the value of the propositions traditionally classified as *future contingents* (following Aristotle) is not correct either. Urquhart interprets Łukasiewicz's values as sets of possible classical values: he sees the third logical value as the set $\{0,1\}$ of the two potential classical values of a future contingent sentence, and the classical values 0 and 1 as $\{0\}$ and $\{1\}$, respectively. The calculation for complex propositions runs according to the classical rules. Thus, e.g., $\{0, 1\} \rightarrow \{0\} = \{0, 1\}$, and the table for implication is the following:

$\rightarrow$	$\{0\}$	$\{0,1\}$	$\{1\}$
$\{0\}$	$\{1\}$	$(1\}$	$\{1\}$
$\{0,1\}$	$\{0,1\}$	$\{0,1\}$	$\{1\}$
$\{1\}$	$\{0\}$	$\{0,1\}$	$\{1\}$

Accordingly, assuming that $\{0\},\{0,1\},\{1\}$ stand respectively for Łukasiewicz's values 0, $^1/_2$ and 1, we get $^1/_2 \rightarrow {}^1/_2 = {}^1/_2$. This is not in agreement with Łukasiewicz's stipulation according to which the output of $^1/_2 \rightarrow {}^1/_2$ is 1. Therefore, Urquhart claims, the original interpretation of $^1/_2$ is incompatible with the concept of *future contingents.*

Among other logical works related to the third logical value, those of J. Łoś and R. Suszko on logical matrices, the theory of structural consequence and, more importantly, Suszko investigations into "non-Fregean" logic, give a new direction to the problem of logical many-valuedness (see Suszko 1957, 1972). Suszko's non-Fregean perspective is characterised by the rejection or, as the author used to say, the "abolition" of the Fregean principle according to which logical values are denotations of propositions – a version of the principle Suszko called the *Fregean Axiom*. This, one could say, is the somewhat idiosyncratic result of Suszko's adopting some aspects of Frege's semantical program while subject to the influence of Wittgenstein's *Tractatus*. While Suszko distinguishes – like Frege – between the content of a sentence and its logical value, he identifies the content of the sentence with the non-classical values. Thus, e.g. the three-valued logic may semantically be interpreted as the logic of sentences admitting three possible contents 0, $^1/_2$ and 1. At the same time – just like Frege – the description of connectives is given using the zero-one valuations (see Suszko 1977). The "zero" and "one" serve then as "genuine" logical values, while Łukasiewicz's three values denote situations – in Wittgenstein's sense – described by sentences. Obviously, the classical part of the construction remains unchanged and the logical values employed are the standard 0 and 1.

The basic logic of the family – sentential calculus with identity (SCI) – formalizes all classical connectives and the "pure" connective of propositional identity, which is specified further in other non-Fregean logics. While equivalence in Suszko's logic expresses the equality of logical values, the identity connective on its part confirms the collapse of the semantic contents of sentences. The fact is that Suszko accepts all other Fregean principles, including the famous *Semantic Scheme*, and builds a family of logics in which some equivalents (i.e., propositions having the same logical values) may not be identical. At some stage of his investigations Suszko attempted a partial formalization of the ontology of the *Tractatus*, which forced him to refer to semantic denotations as situations, and to the situations which obtain as facts. One could thus say that Suszko's interpretation of Łukasiewicz's three-valued logic rests on the distinction between two semantic levels: ontological and logical. While the three "ontological" values – 0, $^1/_2$ and 1 are possible denotations for propositions, at the logical level, two values suffice. The latter is showed by the fact that there is a family of 0–1 functions for the set of formulas which adequately describes Łukasiewicz's connectives.[5] One interesting consequence of this is that, if we follow Suszko, a system of three Łukasiewicz-"situations" is logically two-valued, or bivalent.

In 1922 Łukasiewicz generalized his three-valued construction to the family of many-valued matrices, both finite and infinite, countable and uncountable (see Łukasiewicz 1970, 140). All (sets of) values in these logics can be interpreted in the way Suszko suggested. All finite logics can be reinterpreted as bivalent logics of n situations (see Malinowski 1977). Suszko's interpretation however does not work for infinitely-valued logics.

5.5 General Perspective on Many-Valuedness

The variety of many-valued constructions is quite large (see Malinowski 1993, 2001) and the discussion of their respective motivations and justifications goes beyond the scope of this paper. However, it seems reasonable to focus the remainder of the discussion on the general idea of semantics and the many ways in which many-valuedness has become a familiar fixture of the discipline.

To start with, recall that soon after Łukasiewicz first presented his three-valued system, E. Post's thorough studies of the algebraic properties of truth table semantics led him to introduce a class of finitely many-valued propositional logics. Post's approach was inspired by the very method of truth tables and, in particular, by

[5] More precisely, Suszko presents a solution for Łukasiewicz's negation and implication connectives.

the formalization of the (classical) logic of *Principia Mathematica* (Whitehead, Russell; 1910). The main feature of Post's logics is their functional completeness, which means that they are logics of all possible connectives that may be defined using tables having a given specified number of values. The first general foundations of finitely many-valued semantics and the axiomatic formalization of the corresponding systems were given by Rosser and Turquette (1952). The two-voice discussion in the opening chapter brings a cluster of remarks setting the conditions to be satisfied by logical matrices built on the basis of a set of "logical values", the number of which is – in general – bigger than two. This set is divided into two subsets, *distinguished* and *non-distinguished* elements. Distinguished values play a role similar to the "1" of classical logic and, for this reason, they may be interpreted as different ways of exemplifying truth. By symmetry, the non-distinguished elements represent falsity. In such systems, the rules of interpretation of propositional languages are generalizations of the principles of truth-functionality. Rosser and Turquette posed the conditions that make finitely many-valued logics resemble classical propositional calculus (CPC). This made it possible to simplify the problem of axiomatizing and that of extending many-valued constructions to predicate logic. The general conditions resolve themselves in the above-mentioned principle of extensional interpretation for propositional languages in matrices of the form:

$$M_{n,k} = (U_n, D_k),$$

where U_n is an algebra with operations corresponding to the connectives founded on the set of values $E_n = \{1, 2, \ldots, n\}$, $D_k = \{1, 2, \ldots, k\}$, where $n \geq 2$ is a natural number and $1 \leq k < n$. Assuming, with Rosser and Turquette, that the natural number ordering conveys a decreasing degree of truth, then regardless of the values of n and k, 1 refers to "truth" and n takes the role of "falsity".

In this notation the matrix $M_{3,1}$ of Łukasiewicz's three-valued logic is based on $E_n = \{1, 2, 3\}$ and $D_1 = \{1\}$, while the connectives $\neg$, $\rightarrow$ are defined by the following tables:

x	$\neg x$
1	3
2	2
3	1

$\rightarrow$	1	2	3
1	1	2	3
2	1	1	2
3	1	1	1

Detailed conditions describe the connectives whose occurrence in many-valued logics is methodologically desirable and which are called *standard*. The counterparts of the classical functions appearing there are characterized by the division of the set E_n of logical values into two subsets of designated elements, D_k, and undesignated elements, $E_n - D_k$. Among the "standard connectives" there are those which in the many-valued logics in question play the roles of the negation, implication, disjunction, conjunction and equivalence connectives. Moreover, Rosser and Turquette require that standard logics have, either as primitive or as defined, logical value identifiers, the so-called *j-connectives* which may be interpreted as affirmations and negations. Thus, since the elements of $D_k = \{1, 2, \ldots, k\}$ are values expressing the measure of "truth", the members of the family $j_1, j_2, \ldots, j_n$ of these connectives for $i \hat{I} \{1, 2, \ldots, n\}$ is defined by the condition:

$$j_i(x) \in \{1, 2, \ldots, k\} \text{ if and only if } x = i.$$

The strategy for the axiomatization of many-valued standard logics is based on the use of standard connectives. All classical counterparts are used to prove that an appropriate j-family is definable. Getting an axiom system for a given standard logic is then possible by adopting the standard method of axiomatization for implicative systems. Accordingly, any standard propositional logic is axiomatizable using implication and j-connectives. Implication, for instance, requires three axioms and assumes the rule of *modus ponens*. It assumes further:

$$j_i(\alpha) \to \alpha \text{ for } \iota \in \{1, 2, \ldots, \kappa\}$$

which fixes the quality of values which express the measure of "truth". Thus, a formula that may take only 1, 2, ..., k, as its value, belongs to the system. Two supplementary schemes complete the axiomatization. One of them "translates" the non-classical value order using j's. The other is a kind of universal truth-functionality scheme and stipulates that the value of a formula under an interpretation is a function of the values of all its (propositional) variables.

The Rosser-Turquette style formalization of many-valued logics is very classical in spirit and it applies to the logics which have special connectives appearing as primitives or definables. The presence of the j-connectives makes zero-one description of finite n-valued Łukasiewicz-style logic possible; the Suszko-style reinterpretation was described in Malinowski (1977).

5.6 Sequents and Many-Valuedness

The study of formal proofs inaugurated by Gentzen (1934) and Jaśkowski (1934) for classical and intuitionistic logics was soon adapted to finitely many-valued propositional and predicate calculi. A sequent is a formula of the form $\Gamma \Rightarrow \Delta$ where Γ and Δ are finite sequences, or multisets, of formulas. The sequent means that the conjunction of all formulas of Γ entails the disjunction of all formulas of Δ, in symbols $\wedge\Gamma\Box \vee \Delta$. In the classical case, the entailment, which follows from to the theorem of deduction, is equivalent to $\Box \wedge \Gamma \Rightarrow \vee\Delta$. By virtue of the latter, the sequent receives a truth-functional interpretation: for any valuation, if all the formulas in Γ are true, then at least one formula in Δ is not false. A version of the calculus of sequents having a single formula in place of Δ was used by Gentzen to formalize intuitionistic logic. Schröter (1955) gave a direct generalisation of the classical sequent approach for the many-valued case.

A natural, truth-functional approach to the sequent formalisation of finitely many-valued logics is due to Rousseau (1967).[6] Given a finite $n \geq 2$, the n-valued sequent Γ is an n-tuple:

$$\Gamma_1|\Gamma_2|\ldots|\Gamma_n$$

of finite sequences of formulas. Γ is true under a given interpretation if and only if at least one Γ_i, $i \in \{1, 2, \ldots, n\}$, has a formula which takes the value i. Thus, the components $\Gamma_1|\Gamma_2|\ldots|\Gamma_n$ of Γ correspond to the logical values of the logic under consideration. It is obvious that for $n = 2$, one gets the counterpart of the standard notion of a sequent $\Gamma_1|\Gamma_2$ with its usual truth-falsity interpretation. The key move in the construction of a finitely many-valued sequent calculus consists in accepting an appropriate axiom stating n-valued truth-functionality. To this aim one assumes that for any formula α, the sequent

$$\alpha|\alpha|\ldots|\alpha$$

is an axiom. In addition to this, there are *weakening* rules for every place i,

$$\frac{\Gamma_1 \mid \ldots \mid \Gamma_i \mid \ldots \mid \Gamma_n}{\Gamma_1 \mid \ldots \mid \Gamma_i, \alpha \mid \ldots \mid \Gamma_n}$$

and (i, j) *cut* rules

$$\frac{\Gamma_1 \mid \ldots \mid \Gamma_i, \alpha \mid \ldots \mid \Gamma_n \Delta_1 \mid \ldots \mid \Delta_j, \alpha \mid \ldots \mid \Delta_n}{\Gamma_1, \Delta_1 \mid \ldots \mid \Gamma_n, \Delta_n}$$

for every $i \neq j$, $i, j \in \{1, 2, \ldots, n\}$. The last and the hardest step consists in stating the admissible introduction rules for the connectives and, later, for quantifiers.

An *F-introduction* rule at the position i for an r-argument connective F, has the following form:

$$\frac{\Gamma_1^1, \Delta_1^1 \mid \ldots \mid \Gamma_n^1, \Delta_n^1 \ldots \Gamma_1^j, \Delta_1^j \mid \ldots \mid \Gamma_n^j, \Delta_n^j}{\Gamma_1 \mid \ldots \mid \Gamma_i, F(\alpha_1, \ldots, \alpha_r) \mid \ldots \mid \Gamma_n}$$

where $\Gamma_1 = \Gamma_1^1 \cup \ldots \cup \Gamma_1^j, \ldots, \Gamma_n = \Gamma_n^1 \cup \ldots \cup \Gamma_n^j$, and $\Delta_n^1, \ldots, \Delta_n^j$ are subsets of $\{\alpha_1, \ldots, \alpha_r\}$.

To get an exhaustive description of the connectives in the sequent setting, one has to establish the rules for all positions, i.e. the original values of the logic. This can be done using conjunctive-disjunctive (CD) partial normal forms. However, formulas

[6] Takahashi (1967, 1970) gave a similar formalization.

of a given many-valued logic may have several specific partial CD-forms, which either are defined in the system – as in the "standard" approach – or are "metalinguistic" descriptions of the truth tables. It occurs that a given formula may have several specific partial forms. This implies that establishing the sequent rules is also not unique.[7] Once, we have such a description for the connective, we may write its introduction rules by taking as the premise the set of sequents reflecting the disjuncts, the components of which are positioned at adequate places.

To give an example, let us take Łukasiewicz's three-valued implication as defined in Section 5 above. One may verify that the following formulas are the CD partial normal forms of $p \to q$:

$(p^3 \vee p^2 \vee q^1) \wedge (p^3 \vee q^2 \vee q^1)$ (the first partial *CD* form ; value 1)
$(p^2 \vee p^1) \wedge (p^2 \vee q^2) \wedge (p^1 \vee q^3)$ (the second partial *CD* form; value 2)
$p^1 \wedge q^3$ (the third partial *CD* form ; value 3),

the superscripts 1, 2, 3 when apposed to a propositional variable signify that the variable takes the indicated value. Thus, p^1 reads "*p* takes the value 1", p^2 reads "*p* takes the value 2", p^3 reads "*p* takes the value 3", etc. Accordingly, we get the following three introduction rules for Łukasiewicz's implication:

$$\frac{\Gamma_1, q \mid \Gamma_2, p \mid \Gamma_3, p \qquad \Gamma_1, q \mid \Gamma_2, q \mid \Gamma_3, p}{\Gamma_1, p \to q \mid \Gamma_2 \mid \Gamma_3} \qquad (I \to; 1)$$

$$\frac{\Gamma_1, p \mid \Gamma_2, p \mid \Gamma_3 \qquad \Gamma_1 \mid \Gamma_2, p, q \mid \Gamma_3 \qquad \Gamma_1, p \mid \Gamma_2 \mid \Gamma_3, q}{\Gamma_1 \mid \Gamma_2, p \to q \mid \Gamma_3} \qquad (I \to; 2)$$

$$\frac{\Gamma_1, p \mid \Gamma_2 \mid \Gamma_3 \qquad \Gamma_1 \mid \Gamma_2 \mid \Gamma_3, q}{\Gamma_1 \mid \Gamma_2 \mid \Gamma_3, p \to q} \qquad (I \to; 3)$$

These rules, together with the rules for negation, for the set $\{1, 2, 3\}$ insure a sequent formalization for Łukasiewicz's three-valued propositional logic. To finish with, we remark that the following introduction rules follow directly from the table of negation:

$$\frac{\Gamma_1, p \mid \Gamma_2 \mid \Gamma_3 \qquad \Gamma_1 \mid \Gamma_2, p \mid \Gamma_3 \qquad \Gamma_1 \mid \Gamma_2 \mid \Gamma_3, p}{\Gamma_1 \mid \Gamma_2 \mid \Gamma_3, \neg p \quad \Gamma_1 \mid \Gamma_2, \neg p \mid \Gamma_3 \quad \Gamma_1, \neg p \mid \Gamma_2 \mid \Gamma_3}$$

[7] See Zach (1993) for details.

5.7 Conclusions

Our presentation of the key issues of Łukasiewicz logic and, more generally, of the multiplication of logical values was mainly inspired by philosophical aims. Summing up the discussion, we may say that three conditions are at the heart of the procedures as far as many-valuedness is concerned:

(i) the rejection of some classical laws, or more generally, of classical inferences;
(ii) many-valued "truth functionality", i.e., the extensionality of functions corresponding to the connectives with respect to "logical values";
(iii) the elimination of the "*Fregean Axiom*" identifying logical values with denotations of propositions;

Other properties, such as "compatibility" with classical logic, although apparently helpful for the purpose of grounding semantic or syntactic descriptions, are however not indispensable for actual many-valued constructions.

References

Bolzano, B. (1837), *Wissenschaftslehre*, Sulzbach, Seidel.

Gentzen, G. (1934), 'Untersuchungen über das Logische Schliessen', *Mathematische Zeitschrift*, 39, 176–210.

Gonseth, F. (ed.) (1941), *Les entretiens de Zurich sur les fondements et la méthode des sciences mathématiques 6–9 décembre 1938*, Zurich, S.A. Leemann frères & Cie.

Jaśkowski S. (1934), 'On the rules of suppositions in formal logic', *Studia Logica*, 1, 5–32.

Kotarbiński, T. (1913), 'Zagadnienie istnienia przyszłości' (The problem of existence of the future), *Przegląd Filozoficzny*, VI.1.

Łukasiewicz, J. (1906), 'Analiza i konstrukcja pojęcia przyczyny' (An analysis and a construction of the concept of cause), *Przegląd Filozoficzny*, 9, (2–3), 105–179.

Łukasiewicz, J. (1910), *O zasadzie sprzeczności u Arystotelesa. Studium krytyczne* (On the principle of contradiction), Kraków (PWN, Warszawa 1987).

Łukasiewicz, J. (1913), *Die logischen Grundlagen der Wahrscheinlichkeitsrechnung*, Cracow. Eng. trans., in Borkowski, L. (ed.), *Selected works*, Amsterdam, North-Holland, 16–63.

Łukasiewicz, J. (1920), 'O logice trójwartościowej', *Ruch Filozoficzny*, 5, 170–171. Eng. trans. *On three-valued logic*, Łukasiewicz 1970, 87–88.

Łukasiewicz, J. (1930), 'Philosophische Bemerkungen zu mehrwertigen Systemen des Aussagenkalküls', *Comptes rendus des séances de la Société des Sciences et des Lettres de Varsovie*, Cl. III, 23, 51–77.

Łukasiewicz, J. (1970) *Selected works*, L. Borkowski (ed.). Amsterdam, North-Holland.

Malinowski, G. (1977), 'Classical characterization of n-valued Łukasiewicz calculi', *Reports on Mathematical Logic*, 9, 41–45.

Malinowski, G. (1993), *Many-Valued Logic*, Oxford, Clarendon Press.

Malinowski, G. (2001), 'Many-valued logics', in L. Goble (ed.), *The Blackwell Guide to Philosophical Logic*, Blackwell Publishers, Oxford, 309–335.

Post, E. L. (1920), 'Introduction to a general theory of elementary propositions', *Bulletin of the American Mathematical Society*, 26, 437.

Post, E. L. (1921), 'Introduction to a general theory of elementary propositions', *American Journal of Mathematics*, 43, 163–185.

Rosser, J., Turquette, A. (1952), *Many-Valued Logics*, North-Holland, Amsterdam.
Rousseau, G. (1967), 'Sequents in many-valued logic', *Fundamenta Mathematicae*, 60, 23–33.
Schröter K. (1955) 'Methoden zur Axiomatisierung beliebiger Aussagen- und Prädikatenkalküle' *Zeitschrift für Mathematische Logik und Grunglagen der Mathematik*, 1, 241–251.
Suszko, R. (1957), 'Formalna teoria wartości logicznych' (A formal theory of logical values). *Studia Logica*, VI, 145–320.
Suszko, R. (1972), 'Abolition of the Fregean Axiom, in R. Parikh (ed.), *Logic Colloquium, Symposium on Logic held at Boston, 1972–73. Lecture Notes in Mathematics*, 453, 169–239.
Suszko, R. (1977), 'The Fregean Axiom and Polish mathematical logic in the 1920s', *Studia Logica*, 36/4, 377–380.
Takahashi, R. (1970), 'Many-valued logics of extended Gentzen style II', *Journal of Symbolic Logic*, 35 (231), 493–528.
Urquhart, A. (1986), 'Many-valued logic', in D. Gabbay and F. Guenthner (eds.), *Handbook of Philosophical Logic*, vol. III. Dordrecht, D. Reidel, 71–116.
Zach, R. (1993), *Proof theory of finite-valued logics. Master's thesis*, Institut für Algebra und Discrete Mathematik, TU Wien, 338–353.
Zawirski, Z. (1934), 'Znaczenie logiki wielowartościowej i związek jej z rachunkiem prawdopodobieństwa' (Significance of many-valued logic for cognition and its connection with the calculus of probability), *Przegląd Filozoficzny*, 37, 393–398.

Chapter 6
Leśniewski's Systems and the Aristotelian Model of Science

Arianna Betti

6.1 Leśniewski's Conservatism and Its Source

The systems of Leśniewski, like Frege's, have an unmistakably old-fashioned flavour. They stand to, say, post-Tarskian, post-Gödelian, post-Hilbertian logic like traditional peasant Tuscan bread soup stands to molecular fusion kitchen. Why is that?

According to suggestions recently put forward, which rely on van Heijenoort's opposition "Logic as Language vs. Logic as Calculus", or similar dichotomies, Leśniewski's attitude to logic was similar to Frege's insofar as it matched Frege's "Logic as Language" rather than Boole's and Schröder's "Logic as Calculus".[1] What grounds the old-fashioned aura of Leśniewski's systems, so goes the suggestion, is Leśniewski's adherence to the "Logic as Language" paradigm.

Is this correct? In introducing his opposition, van Heijenoort builds on a remark by Frege on the *Begriffsschrift* as a system embodying two Leibnizian ideals that are in fact not opposed: *lingua characteristica* and *calculus ratiocinator* (van Heijenoort 1967: 233). But van Heijenoort's dichotomy remains very sketchy, so sketchy that one does not seem to get very far by applying it. Two things can be done to save its gist. One is beefing it up. This was done by Jaakko Hintikka in his refurbished *Language as Universal Medium* vs. *Language as Calculus*. The other option is tracing the source of van Heijenoort's opposition in the history of philosophy, and go back, if at all possible, where it all started. In this paper I shall go for the latter. I shall leave for another occasion an account of why I think this is a much more fruitful option than Hintikka's. For my purposes here it will suffice to show that there is another way to account for Leśniewski's conservatism, a way that makes appeal to a millennia-old recipe for building proper deductive systems: a venerable model of scientific rationality to which I will refer in what follows as The Aristotelian Model of Science. The main purpose of this paper is to illustrate, then, that Leśniewski's systems follow this model closely.

A. Betti (✉)
Vrije Universiteit Amsterdam, Amsterdam, The Netherlands
e-mail: ariannabetti@gmail.com

[1] Cf. Sundholm (2003: 113), whom I follow in my (2004).

S. Lapointe et al. (eds.), *The Golden Age of Polish Philosophy*, Logic, Epistemology, and the Unity of Science 16, DOI 10.1007/978-90-481-2401-5_7,

The significance of Leśniewski's adherence to the Aristotelian Model can be properly gathered on the basis of three facts. First, there is a close link between Leibniz's project of a *lingua characteristica*-cum-*calculus ratiocinator* and the structure of a deductive science according to the Aristotelian Model of Science as presented in Section 2. The two ideals indeed show their full meaning if portrayed against the background of the Aristotelian Model, which was dominant in Leibniz's times. Secondly, Frege was a close follower of the Aristotelian Model – and in a way which is relevant to his ideal of a *Begriffsschrift* as embodying Leibniz's project; thirdly, Leśniewski, in turn, wholeheartedly took up this aspect of Frege's project. Of these three facts only the second has been already described in some detail.[2] The first, the link between Leibniz and the Aristotelian Model, I will have to take largely for granted: I will offer very little evidence for it, and however telling and interesting that evidence might be, the point will be by no means sufficiently argued for. As to the third, i.e., Leśniewski's project as being substantially Frege's, this paper is an important step to establishing this very point.[3]

It is no doubt important that both Frege and Leśniewski adhered to the Aristotelian Model closely. No scholars interested in the foundational debate, at least from the logicist side, should overlook this. Leśniewski's work was the most impressive and accomplished logicist attempt in the footsteps of Frege, and the extent to which both Leśniewski and Frege follow the Aristotelian Model should make us suspect that it is the most promising framework to answer questions like the one we started from, that is, what it is that makes the systems of Leśniewski so distinctively old (that is, old like Good old *ribollita*). And if the Aristotelian Model is the core of Leśniewski's conservatism, then the suggestion becomes likely that van Heijenoort's or similar frameworks apply to Leśniewski to the extent that they can be subsumed under the Aristotelian Model. But whether that is what we should conclude, as already mentioned, will be dealt with on another occasion.

6.2 General Framework

Between 1919 and his death in 1939, Stanisław Leśniewski developed a nominalistic system of the foundations of mathematics split into three axiomatic, fully extensional, hierarchically structured, deductive theories: Protothetic, Ontology and Mereology.[4] Protothetic is a quantified propositional calculus with variable functors. Ontology is a calculus of names and corresponds roughly to predicate calculus

---

[2] Cf. De Jong (1996), De Jong and Korte (forthcoming), who follows for Frege a strategy similar to the one followed here for Leśniewski, that of invoking the Aristotelian or (Classical) Model of Science rather than van Heijenoort's framework to understand Frege's attitude towards logic as a science and logical language.

[3] Puzzling and sad, but true, the influence of Frege's logical work upon Leśniewski has never been seriously investigated. For a preliminary recognition, see Woleński (2004b).

[4] I will not go here into the formal characteristics of Leśniewski's systems in any detail. For this see LeBlanc (1991), Luschei (1962), Miéville (1984).

with identity. Mereology is a theory of parts and wholes. Mereology presupposes Ontology, which in turn presupposes Protothetics. These theories, as I shall show in the following sections, conform closely to the Aristotelian Model or, for the sake of brevity, "the Model", which reflects the epistemological position known as classical foundationalism in the philosophy and the methodology of science.

Let us call a (deductive) system *S* an Aristotelian science iff:

(I) (*domain postulate*) Every proposition and each term (or concept) of *S* concerns one and the same specific domain of being.

(IIa) There are in *S* a (finite) number of so-called fundamental terms (concepts), so clear in meaning as to require no further explanation.

(IIb) (*composition postulate*) Any other term occurring in *S* is composed of (or is definable from) these fundamental terms (concepts).

(IIIa) There are in *S* a (finite) number of propositions called fundamental propositions.

(IIIb) (*proof postulate*) All other propositions of S follow (are grounded in, are provable) from the fundamental propositions.

(IV) (*truth postulate*) Every proposition of *S* is true (and known to be such).

(V) (*necessity postulate*) Every proposition of *S* is necessary (in some sense or another).

This particular systematization in five conditions is adapted from De Jong (2001: 330), which builds on similar ones set up among others by Beth (1959: 32) which in turn followed Scholz's characterization (1975: 52), the first and perhaps best known attempt at a systematization.[5]

The Model was enormously influential for more than two millennia. It dominated in particular the philosophy of science of the Sixteenth, Seventeenth and Eighteenth centuries, counting among its prominent followers Newton, Pascal, Spinoza, Descartes, Leibniz, Wolff, Kant, and it still played a crucial role later in Bolzano, Husserl, Frege and, as we shall see, Leśniewski.

In the history of philosophy, however, attempts at formulating explicitly the ideal of science epitomized in the Model are rare (one exception is the eight rules of scientific method in the Logic of Port-Royal), and no mention of the Model itself is to be found in any of the philosophers mentioned above. The Model is thus a reconstruction *a posteriori* of the way in which philosophers have traditionally thought about what a proper science and its methodology should be.

---

[5] For this Section and Section 4 I am much indebted to Wim de Jong. For more information on the Model, its historiography and the differences between various attempts at encoding the same ideal of science, see De Jong and Betti (forthcoming), where an updated and enlarged set of conditions is proposed. This paper does not use the updated version of the Model, but the one available in print already for some years. The updated Model is used in Betti (forthcoming), of which parts of the present paper are a further elaboration. Betti (forthcoming) has a different focus; furthermore, it contains information on Leśniewski's systems absent here and a different discussion of the various conditions of the Model.

According to the Model, a proper (an "Aristotelian") science has the structure of a more or less strictly axiomatized system with a clear operative distinction between fundamental elements – also called *principles* – and non-fundamental elements. The sense in which a science in the sense of the Model is an "axiomatized system" is broader than the one the expression usually has. One way to put this is to say that the Model combines both axiomatics proper and the so-called regressive method of conceptual analysis (and synthesis, cf. Beaney 2003), insofar as it leaves room both for a system of propositions ordered by relations of grounding (or, on a weaker understanding, deducibility) and for a system of terms ordered by means of definitions.

The philosophical debate around the Model before Frege and Leśniewski is of great relevance both to understand their choices and to clarify some issues in the reception of their ideas that would otherwise remain obscure. Consider the two conditions of the Model that have most often given rise to divergences in history, condition I and V. As we will see, those are also the most difficult conditions to assess for Leśniewski's systems. Condition I, the domain postulate, is linked to the debate on the status of logic as a science, and logicism represents a particular take on the place of logic in the hierarchy of sciences, while condition V, the necessity postulate, is linked to the age-old problem of proving the principles, which famously opposed Descartes and Leibniz.[6]

In the following sections I will show how Leśniewski's systems fulfil the conditions set out above, starting from the less troublesome Composition and Proof Postulates (Section 3), and following with Domain (Section 4), and Truth and Necessity Postulates (Section 5), all three of which pose more problems. As this is a first attempt to relate Leśniewski to the Aristotelian Model of Science, I shall concentrate on some aspects at the expense of others.

6.3 Axioms and Primitive Terms in Leśniewski

In accordance with condition IIIa, all three Leśniewskian systems have a finite number of fundamental propositions (axioms). Leśniewski elaborated (and implemented) a few "aesthetic" requirements for well-constructed axiomatic theories, linked to conditions IIIa and IIIb, both for any axiomatic theory whatsoever and for his own theories.

Some general requirements on the axioms of a theory (the *axiom system*) are "quite trivial and known to everybody" (Sobociński 1955/56: 56[7]), for instance

---

[6] In particular Leibniz's take on the question is of importance in trying to understand Schröder's criticism of Frege's *Begriffsschrift* as *lingua characteristica* in Leibniz's sense. Cf. Korte (forthcoming: 6). And this, in turn, is important in order to understand Leśniewski's project as well.

[7] Leśniewski discussed with his students the requirements of a well-constructed axiom system, but never published anything on the subject himself. Our most important source for Leśniewski's view on the subject is Sobociński (1955/56).

consistency (of the axioms, not of the theory), equivalent to the requirement that all theses singled out as axioms belong to the field of the theory and that the rules of procedure be adjusted to the primitive terms occurring in the axioms. Another "quite trivial" requirement is that the axioms be *mutually independent*, that is, it should be impossible to derive axioms from one another following the rules of procedure of the theory (Sobociński 1955/56: 56–7).

Leśniewski also formulated some special requirements on the axioms of a theory. The *length of the axiom system* should be the *shortest possible*, and in it the *lowest possible number of variables* should occur (Sobociński 1955/56: 62 a, b; Leśniewski 1939: 23, English translation 1991: 671).

According to Leśniewski's classification of types of axioms, there are axioms *of existence*, axioms *resulting from the rules of procedure* and axioms *proper*. An axiom of existence is an axiom assuming the existence of certain objects, like for instance the axiom of infinity. The theses of Leśniewski's systems that concern extensionality can be seen as examples of axioms resulting from the rules of procedure. An axiom proper is a thesis placed at the beginning of a system as a starting point. From an axiom proper one cannot deduce that something exists.

According to Leśniewski, a well-constructed axiom system should consist, if possible, of a single axiom proper (Sobociński 1955/56: 60), and for him this was a very important requirement. Leśniewski did not consider axioms resulting from rules of procedure to be acceptable in an axiom system because there are infinitely many of them resulting from these rules or procedure. Axioms of existence are not acceptable either since a deductive theory (in particular a logical theory) must be for Leśniewski *ontologically neutral*, that is, it should not come with presuppositions as regards the sort of objects there are in the world or their number. This is an important characteristic of Leśniewski's system to which I shall come back in Section 5 below.

Among the requirements applying to Leśniewski's systems in particular are *canonicity*, *ontological uniformity* and the *internal independence of single axioms*. An axiom system is *canonical* if a) it contains a single axiom which b) is based on a single primitive term; c) has the form "of an equivalence whose left-hand side contains only the simplest possible sentence which includes the primitive term" (LeBlanc 1983: 101); d) the main quantifier "[binds] the variables appearing in the expression on the left-hand side of the equivalence and no others" (Ibid., cf. also Leśniewski 1929: 332). The requirement concerning ontological uniformity applies to all theories presupposing Ontology and says that "although we are allowed to use any protothetical constants for the purpose of constructing axiom systems, the only ontological functor that can be used in this connexion is the one which plays the part of the primitive term in the accepted axiom of ontology" (Sobociński 1955/56: 62). The requirement that the axioms be internally independent does not apply to Protothetic and "stipulates that all the theses which can be derived from a single canonic axiom in virtue of the logic of propositions, or the laws concerning the use of the quantifiers, should be mutually independent" (Sobociński 1955/56: 63).

Leśniewski adopts the requirement which is also "known to everybody" that the axiom system be adequate (Sobociński 1955/56: 56–7), that is, the requirement that all true and desirable theses belonging to the theory ought to be derivable from the axiom system. This corresponds to condition IIIb in the Model. Another additional Leśniewskian requirement that relates to IIIb is that the axioms be organic, that is, they should contain "no segment which is a thesis or becomes one as soon as its variables have been bound with an appropriate quantifier" (Sobociński 1955/56: 60). Leśniewski considered this to be highly significant since the organicity requirement has bearings on the length of the axiom system.[8]

As for primitive terms and the composition Postulate (IIa) every Leśniewskian deductive theory has a finite number of primitive terms. But Leśniewski strengthened this condition further. First, following Peano's collaborator Alessandro Padoa, Leśniewski required that *the primitive terms be mutually independent*, that is, he demanded that, in analogy with to the independence of the axiom system, in axiom systems in which several primitive terms occur, "none of these [primitive] terms can be defined with the aid of the remaining ones". Leśniewski also stressed that the primitive terms have to be simple. A precise definition of the simplicity of primitive terms was given by Adolf Lindenbaum in 1936 following Leśniewski's ideas.[9]

Secondly, Leśniewski deemed strongly desirable, albeit not absolutely required, that an axiom system be based on one *single* primitive term. To see the exquisite refinement to which this desideratum was brought by Leśniewski, note that systems of Protothetic can be based on axioms that contain double implication as their single primitive term.[10]

Now, the double implication is a particularly desirable primitive for two basic reasons. First of all, in a Leśniewskian system definitions are formulated with the aid of the double implication and are, like theorems, (true) *theses* belonging to the system, not metalinguistic abbreviations. It is completely in keeping with the Model that, if definitions belong to the system and we have a single primitive term from which every other are defined, we *cannot* introduce a special term like '$=_{\mathrm{df}}$' with the sole purpose of formulating definitions (Leśniewski 1929: 11; 1991: 418). Definitions must be formulated with the aid of the single primitive. That Leśniewski, thanks to a result by Tarski (1923), was able to construct systems of

---

[8] See for more requirements Sobociński (1955/56: 63), and LeBlanc (1983: 101; 109–110). Both works have been fundamental sources for this section.

[9] In the pioneering Lindenbaum (1936). On simplicity cf. Sobociński (1955/56: 58–9).

[10] Note that equivalent systems of Protothetic, Ontology and Mereology based on different primitives are possible. If one mentions only *single* axiom systems: for instance, Protothetic can be built on implication (Leśniewski 1929: 58; 1991: 467); Mereology on *pt* (proper part), *el* (element), *extr* (outside of), *ov* (overlapper), *ctn* (container), *Kl* (complete collection), etc., cf. LeBlanc (1983: 100*ff*; 115). This means that Leśniewski's version of the Model is not absolute in the sense that axiomatic structures are a Mirror of Nature whose primitives are discovered (for they are rather chosen) and the order of their theses immutable. Leśniewski did favour some systems above others, according to the desiderata spelled out in this section, but among those criteria none follows metaphysical primacy. All this has bearing on condition V, the Necessity Postulate insofar as one way to construe this condition in history has been to see axiomatic structures as mirroring nature in the sense mentioned above. Thus, Leśniewski's systems cannot be said to obey to the Necessity Postulate in this sense. Moreover, Leśniewskian definitions by no means reveal the essence of the thing defined and cannot be linked to any notion of necessity in an ontological sense. The sense in which Leśniewski's systems can be said to follow condition V is explained in Section 5.

Protothetics based on double implication as the single primitive is an achievement of remarkable elegance. The second reason is that axioms also have the form of equivalences of a certain structure, so the single primitive term of Protothetic also is (or more correctly, can be) the only constant functor appearing in the axiom system of Protothetic.

Every non-primitive term of Protothetic, Ontology and Mereology is definable by means of its primitive terms (and rules) alone. This agrees with condition IIb in the Model, the composition postulate. Note that in a Leśniewskian system, the first rule of method in the Logic of Port-Royal, "use no term whose meaning had not previously been clearly explained",[11] becomes a technical requirement: explaining the meaning of a term amounts to adding to the system a definition of the term in question. Before the relevant definition is added, the term cannot be used, for it is meaningless: the term acquires its meaning via its definition.

Primitive terms are defined as "constants which occur in the axiom system and do not belong to any of the presupposed theories" (Sobociński 1956: 57). Not only must the primitive terms be chosen in order to grant an adequate axiom system, they must also be *exclusive*. The latter requirement stipulates that no defined term should be used in the axiom system, and the requirement is a formal one as it is secured by Leśniewski's rules of procedure.

The fact that Leśniewski formulated explicitly separate sets of requirements for terms and axioms matches quite closely the double ordering of propositions (IIIa-b) and terms (IIa-b) in the Model.

6.4 The Domains of Leśniewski's Systems

In what sense do Protothetics, Ontology and Mereology follow condition I of the Aristotelian Model of Science, the Domain Postulate? The question is not easy to answer, for condition I itself comes with two difficulties. First, it is clear enough that its function is to insure that sciences possess an internal unity on the basis of which they can be delimited. But difficulties arise when we wish to get a grasp of the domain of a science less loosely characterized than this. Not only is there no agreement as to what Aristotle meant exactly by "domain", there also seems to be no sufficiently clear modern alternative notion.[12] Secondly, and this is connected to the first difficulty, there have been in history two fundamental interpretations of the notion of domain, which condition I of the Model can accommodate. Let us call them the *term-interpretation* and the *object-interpretation*. The object-interpretation sees the highest specific genus of a science as a collection of *objects*: a science investigates the attributes of certain objects, whose existence is assumed, and which form its subject-matter. The term-interpretation takes the genus to be a collection of

---

[11] Quoted from De Jong (1996: 299 n. 15).

[12] And to make things worse, history seems to have known of no better solutions to this problem: Bolzano found the notion intricate, and so did Husserl. For Bolzano see De Jong (2001, in particular 332 n. 10) for the references to Bolzano I mention in the next page. For Husserl cf. (1900/01), *Prolegomena*, §2.

certain *terms*, that is, expressions with a certain meaning of which the language in which a science is formulated is composed, and which delimit the subject-matter of that science. Can we get a more precise idea of how the domains of Leśniewski's theories are individuated? And which interpretation is suitable for them, the object-interpretation or the term-interpretation?

As for the first problem, the notion of domain and its application to Leśniewski's systems, note that one should avoid thinking of the domain as a special *set* of objects unless this is done in a rather generic sense, and not in the modern mathematical sense. This might strike one as excessively cautious: it seems perfectly natural to speak of, say, the set of integers as the domain of arithmetic. But this is just one step away from construing the domain of a science as the modern technical notion of a (model-theoretical) domain as the universe of discourse, whether or not the latter is identical with the notion of the range of quantification of variables. And the temptation of taking this step we must resist. For one thing, the terms mentioned in the Model are not meaningless inscriptions, graphic manifestations in need of interpretation in a model-theoretical sense. By "terms", meaningful (categorematic, as a rule) expressions are meant. In general, Leśniewski's work – like Frege's – cannot be squeezed in a model-theoretical framework. To see why this warning is important, consider the fact that so far the few analyses connecting the work of logicians like Frege and Leśniewski to the "Aristotelian conception of axioms and proofs" refer to Scholz (1975) as an authoritative source for this conception (Sundholm 2003: 123 Note 9). Scholz saw the domain of a science as "the totality of the arguments whose substitution for the free variables in a formalised axiom-system turns these axioms into true sentences" (1975: 62). This is not what the domain of a science can be said to be, neither in the Aristotelian ideal nor in the realized axiomatics of Frege's or Leśniewski's. Neither Frege, nor Leśniewski allowed free variables in their systems (Leśniewski 1929: 31; English translation 1991: 439). And most importantly, as Mignucci already noted, if Scholz were right on the notion of domain to be associated with the Aristotelian conception captured in the Model, axioms would be empty schemata, and not true propositions (Mignucci 1975: 192). This would unacceptably contradict the Truth Postulate at the outset, which both Frege and Leśniewski accepted.

As to the second problem, if one asks what the specific domains of Leśniewski's Protothetic, Ontology and Mereology are, one might be tempted to answer that the domains of the sciences in question are that of propositions, that of names, and that of objects that are ingredients of other objects. This, however, would suggest that the criterion used to fix the domain of various sciences is not homogeneous across sciences, as it oscillates between the object-interpretation for Mereology and the term-interpretation for Protothetics and Ontology. I wish to maintain that for Leśniewski's systems the term-interpretation is the appropriate one.[13]

[13] According to Jonathan Barnes (1999: 119), Aristotle himself chose the term-interpretation. A supporter of the latter who is more directly relevant to Leśniewski from the historical point of view is Bolzano (I conclude this on the basis of §1, §393.2B, §536).

At first glance there is a number of possible difficulties with the term-interpretation of the domains of Leśniewski's systems. Before we see how we can dispel these difficulties, I will give several reasons why the object-interpretation cannot do.

Let us first consider the suggestion that the domain of the science is singled out by the primitive terms appearing in the axioms of the science in question (Scholz 1975: 62). Traditionally, the choice of primitive terms seems to have been limited to categorematic terms.[14] But in a Leśniewskian system primitive terms are syncategorematic: typically, the equivalence functor for Protothetic, that is, a proposition-forming functor with (two) propositional arguments (s/ss), the copula for Ontology, that is, a proposition-forming functor with (two) nominal arguments (s/nn), and the functor "part of" for Mereology, a name-forming functor with (one) nominal argument (n/n). What follows from this, on the object-interpretation? The domain of Protothetic, Ontology and Mereology would be given by objects that are the ontological counterparts of functors like "$\leftrightarrow$", "ε", and "$<$", presumably particular relations. No matter how hugely improbable this is in itself, it cannot be right at any rate for Leśniewski because by the time he built his systems he granted no place to relations in the world (Lesniewski 1927: 183 Note 1; English translation 1991: 198 Note 6).

Secondly, take the suggestion that what is proved and what is unproved in a science must bear on the same domain (Scholz 1975: 53 *An. Post.* 87 b 1 *ff*). This seems right. But now consider the fact that in Protothetic a constant for the false sentence can be defined, and in Ontology a constant for the empty term. A false sentence and an empty term do not have ontological counterparts by definition (this is not to every philosopher's liking, but what matters of course here is Leśniewski's). If we stick to the object-interpretation, then both Protothetics and Ontology would be just at odds with the Domain Postulate.

Note, thirdly, that it would be completely out of line with Leśniewski's convictions to think of the domain of Protothetics as a collection of special objects like states of affairs.

In favour of the object-interpretation one may object that some Leśniewskian passages seem to support it. For instance, Leśniewski claimed to have formulated in Ontology a certain kind of "general principles of being", and both Kotarbiński (quoted by Leśniewski himself) and Ajdukiewicz compared Ontology to Aristotle's metaphysics as "first philosophy" and considered Ontology as a science taking up the role of Aristotle's metaphysics as a general theory of objects (Lesniewski 1931: 161–163, English translation 1991: 372–4). However, in the passages in question Leśniewski took Ontology to be correctly described as a calculus of names. And moreover, when discussing Ontology some modern interpreters cautiously say that it is a theory of being in the quite innocuous sense of a theory of the copula "is" (Küng 1981: 170). No doubt the theory allowed by the directives agrees with

[14] Se for instance Kambartel 1975: 217, who quotes concepts as thinking, movement, light (like in Locke), time (in Pascal), etc.

Leśniewski's nominalist intuitions, but thanks to its extreme generality Ontology is by no means ontology understood as a description of what there is. It is a Leśniewskian theory of being as much as syllogistic was an Aristotelian theory of being, that is, not quite.

Let us now turn to the term-interpretation. As I said, the term-interpretation has problematic aspects but as I will try to show, these aspects do not give us reasons to abandon it. The first problem is this: saying that the domain of Ontology, as a calculus of names, and as a proper part of logic, consists of *names*, even if this is motivated by an analogy with Protothetic as a calculus of propositions, seems to commit us to saying that Ontology is the science of all names, and this does not seem right. Ontology says things about objects, not about names. When read in ordinary language theses of Ontology say, for example, things like

(1) no object is contradictory,
(2) *x*s exist if some *y* is *x*.

In (1) and (2), it seems apparent, we are talking about objects, objects that are *x*, for instance, not about their names. In a sense this is correct. There is a sense, however, in which it is not, that is, the sense according to which Ontology would commit us to the presence in the universe of (the referent) of every *x, y, z*... appearing in its theses. It is not difficult to see this. As mentioned, Ontology allows empty terms, one can thus state in Ontology theses that, in ordinary language, read like this:

(3) A machine under 200 euros that gets you foam-topped orange juice does not exist.

(3) is true. "A machine under 200 euros that gets you foam-topped orange juice" is an empty term, so no machine of that kind is present in the universe – well, that's why (3) is true (of course, this is by no means plain for a number of philosophers). But this also means that (3) is about such a machine only in the sense that (3) *can be stated*. I should note at this point that the way in which I put things just now is an oversimplification of how things actually work in Ontology and its language.[15] Still the point can be made, albeit on this oversimplified basis, that the answer to the question: What do theses of Ontology say? provides no argument against the term-interpretation.

A second problematic aspect of the term-interpretation is the following. Suppose we accept that the domains of Protothetics, Ontology and Mereology must be construed in a way that does justice to the term-interpretation. How does this work? We saw that it would not be convincing to say, for instance, that the domain of Ontology is that of names. What is the alternative, though? One could think

[15] See on this Lejewski (1954). The question is both an interesting and a complicated one because it touches on how we should understand quantifiers in Leśniewski's systems, but no one has explained this in a satisfactory way yet. Such an analysis could, I think, be offered rewardingly in a discussion involving the Domain Postulate, and to be really convincing and exhaustive it should also involve the development of the notion of variable from Bolzano onwards. This would go, however, far beyond the aims of this paper.

of linking the subject-matter restriction to the restriction concerning the kind of semantic categories of the terms that are allowed in the formal systems. So, roughly, every proposition of Protothetic contains only protothetical terms, that is, terms of protothetical semantic category. In Ontology categories of a new sort are added, namely ontological categories. But now how about Mereology? In Mereology no new kind of semantic category is introduced, that is, there are no specifically mereological semantic categories. So it seems that the term-interpretation so far did not lead us to a satisfying criterion to delimit sciences. But we could still try to argue as follows: the domain of a science is given by the primitive terms and by *the sort of (constant) terms that can be defined* once the primitive terms are given. In this sense, thus, the domain of Leśniewski's Protothetic is fixed once we say that it is a collection of truths whose constants are either propositional constants (like 0 or 1) or proposition-forming functors of propositional arguments (like the propositional negation "∼": not). The domain of Ontology is fixed once we say that it is a collection of truths among whose constants are, in addition to Protothetical ones, also ontological constants (like V: object), proposition-forming functors of ontological arguments (like "ex": exists), name-forming functors of ontological arguments (like the nominal negation "¬", non-). What we can say about Mereology, now, is that the domain of Mereology is fixed once we say that it is a collection of truths among whose constants we also find, in addition to Ontological and Protothetical ones, other constants which are all name-forming functors of ontological arguments. Mereology is a collection of truths in which propositions of structure "S is *mr*<P>" where "*mr*" is a mereological functor are allowed to appear.

This conclusion about Mereology at first looks stunningly uninformative, and indeed there seems to be no better conclusion. But that there is no better conclusion is less surprising once we consider two relevant and interrelated issues connected traditionally to the Domain Postulate: first, the hierarchy of sciences, or *subalternation*, which is traditionally linked to discussions of Aristotle's prohibition-rule on kind-crossing mentioned above, that is, the appeal in a science *S* to principles laying outside *S*, in particular the appeal to propositions not belonging to *S* to prove propositions of *S*; secondly, the old problem of whether disciplines like logic and metaphysics should or should not be considered Aristotelian sciences in their own right. Leśniewski's systems form a hierarchic structure in which the first two grades are occupied by the two Aristotelian sciences that form, taken together, logic. Our problem of finding a homogeneous criterion is linked with this hierarchical disposition, and it is complicated by treating logic as an Aristotelian science.

The reason for the difficulty lies thus in the chasm between Protothetic and Ontology taken together, on the one hand, and Mereology on the other. The chasm is the one between the logical and the non-logical and surfaced in history when logic began to be considered an Aristotelian science that would serve as a foundation to all other sciences. We can come to no satisfying solution to the problem of the application of the Domain postulate if we do not take this into account. While a distinction between the specific domain of Protothetic and that of Ontology can be given in an absolutely formal way, that is, on the basis of the semantic categories of the terms involved, this cannot be done for Mereology. The difference between two non-logical theories as to their domains must be given in some other way than by logical grammar.

It would be interesting to try to determine when the domains of two non logical theories are given in a sufficiently precise manner. But for the aims of this paper the conclusion suffices that if we treat logic as an Aristotelian science the Domain postulate in the term-interpretation still applies, in other words that there is not ground for saying, for instance, that we must switch from the term-interpretation to the object-interpretation of domain in the case of Mereology.[16]

6.5 Leśniewskian Systems as Systems of Truth

It is doubtless that for Leśniewski axioms are true. Unfortunately, he seldom says so. One rare explicit passage about this, from the first axiomatization of the "General Theory of Sets" of 1916 – Leśniewski's Mereology-to-be – is the following:

> The psychological 'source' of my axioms is my '*intuitions*', which means simply that I *believe* in the truth of my axioms, but *why* I believe I am not able to tell, for I am not acquainted with the theory of causality
>
> (Leśniewski 1916: 6; English translation 1991: 130, amended).

It is seldom safe to draw promiscuously from both early and later Leśniewskian writings. This, however, is one of those safe times: Leśniewski never changed his mind about the truth of axioms.[17]

In an Aristotelian science, conditions IIIb and IV taken together imply that *truth follows from truth*: axioms are true, and every other proposition that can be proved in the theory from the axioms by following the derivation rules at our disposal, and every other thesis that can be introduced as a definition, is also true. The reason why Leśniewski wanted to state his formal methodology as precisely as he could – something he actually managed to do with unmatched precision – was expressly that of building a paradox-free system of the foundations of mathematics, which is at the same time also a properly built system of truths. For, in a properly built system, from true premises an antinomy cannot follow by valid rules. Leśniewski was obsessed with blocking antinomies by disambiguating rules which would let truth follow from truth not just in any way, but in an *intuitively* valid way, that is, on

[16] A problem might very well arise if we were faced with a switch of criteria between extra-logical theories, that is, Mereology and the sciences subordinated to it. But this issue lies beyond the scope of this paper insofar as Leśniewski did not build any other extra-logical theory than Mereology. The information we have on Lejewski's Chronology – a LeŚniewski-style theory of objects in time to follow Mereology – is insufficient to provide a counterexample to what I say here, cf. the 'Aristotelian' Lejewski (1982).

[17] Cf. Leśniewski (1929: 6; Eng. Tr.: 413). This is an even earlier quote: "I call a systematized aggregate of [...] true propositions [...] a science. [...] The aim of a consistently carried out scientific criticism is the exclusion [...] of all false propositions [...] from the system of propositions which is considered as a science." Leśniewski (1912: §*14*, 1991: 35, amended)

one reading of "intuitively valid", in an epistemically acceptable way.[18] Reaching this aim has been Leśniewski's major contribution to the axiomatics of deductive systems.

As said above, the Model codifies a foundationalist analysis of knowledge. This is clear if we take conditions III, IV and V together. Axioms are, on one account, self-evident, or at least, on a weaker reading, they are accepted without proof within the system itself of which they are axioms. Such is, once again, the thinking behind the 1916 proto-Mereology:

> Now, I treat my system expressly as a hypothetico-deductive system, from which follows that, actually, I assert only that from the propositions I call 'axioms' follow propositions which I call 'theorems'. [. . .] My axioms don't have a logical 'source', which means simply that these axioms do not possess a proof in my system, like no other axiom whatsoever is, in the nature of things, proved in the system for which it is an axiom
> (Leśniewski 1916: 6; English translation 1991: 130, amended).

Now, according to the traditional analysis of knowledge, believing some truth *p* *is* tantamount to knowing *p*, if the belief that *p* is justified. *p* can be justified either immediately or mediately. In the Model, III, that is, the proof postulate implies that the truths that follow from the axioms have a mediate epistemic justification. What about the axioms? Well, since they are not proved in the system, it seems that they can only be immediately justified. But are they?

By Leśniewski's time this issue had tormented philosophers of science for millennia. Leśniewski refrains from declaring his axioms self-evident; he seems to think of them as having far weaker epistemic grounds than self-evidence. Now, one traditional way to argue for a direct grasp of the axioms had been to appeal to special epistemic faculties or sources of knowledge, like intuition (Scholz 1975: 58; De Jong 1996: 311), and the above quoted passage might appear to imply something of the sort. But the use of "intuition" there is not technical despite the fact that in the few places where Leśniewski talks about the status of the axioms and the rules of his systems, he resorts to this terminology.[19] His triple remark: axioms are true, their epistemic ground comes from a common-sense notion of intuition, axioms are unprovable logically, was meant, as I take it, to keep philosophers quiet by just mentioning the problem while refusing to go into the matter. Yet if Leśniewski does not intend to go into the matter, this also means that Leśniewski is steeped in the very same foundationalist tradition as those he wants to keep quiet. And Leśniewski's triple remark means also that he knows no better solution to the traditional problem of proving the principles.[20]

As to condition V, the necessity postulate, if all propositions of the system must be necessary, the axioms also have to be necessary. But if this means that they have to be self-evident, as we have just seen, it does not seem easy to accommodate this

[18] Cf. Leśniewski (1929: 6; Eng. Tr.: 413).

[19] Cf. Leśniewski (1929: 6; Eng. Tr.: 413).

[20] Husserl's essences-solution in the *Prolegomena* (cf. Philipse 2004 especially Section 6 and Philipse 1983, especially pp. 120–1) would not do for the nominalist Leśniewski, and neither would Frege's contorsions (cf. De Pierris 1988).

condition. Nevertheless, the question of the self-evidence of the axioms seems by no means settled in the above passages from 1916. Leśniewski's later requirements concerning axioms and primitive terms for his logical systems, which we saw in Section 3, seem indeed intended to fulfill condition V in the following sense: axioms are self-evident because the evidence for their truth rests upon nothing else than the meanings of the expressions used in them. This is supported by the following. First, Leśniewski's requirements related to conditions II and III prescribe that axioms should have the highest informative value yet the simplest, most elegant and most transparent form: the primitive term should be a single one of the lowest grammatical degree placed in the initial segment of the shortest possible single axiom with the form of an equivalence with the lowest possible amount of variables. Secondly, largely as a consequence of the latter, an axiom in Leśniewski is, we could say, a *self-definition* of the primitive term. Thirdly, the primitive terms always comply with the second part of IIa, that is, they are always intuitively clear, and in their ordinary use, in a pre-theoretical way, in no need of further explanation: "if and only if", "is", "part of". This is as close as we get to a solution to the problem of proving the principles. For what can the idea behind all this be, if not that of finding the most adequate way to present the mind with all the information needed to get a grip on the axioms? It is important to stress that while this might well have the consequence of reducing, on the side of the reader, a certain kind of theoretical effort to a minimum, it does not mean that understanding axioms like Leśniewski's is supposed to be an easy task.[21] It would be quite strange to be able to understand at a glance extremely general, compact and difficult statements like the following axiom of Protothetic:

$$\mathrm{PA}\forall pq((p \leftrightarrow q) \leftrightarrow (\forall F(F(pF(p(\forall r(r))) \leftrightarrow (\forall s(Fqs) \leftrightarrow (q \leftrightarrow p))))))$$

Nonetheless, every part of it is, in itself, perfectly understandable and (especially if one were to write it in Leśniewski's own symbolism) unambiguous.

An important point in this connection concerns, once again, the extreme care with which Leśniewski formulated the rules for his systems in his maniacal attempts to exclude error. The way in which this is accomplished is in some respect similar to the way in which Leśniewski approaches the problem of proving the principles: supplying all information to support the mind, this time in its inferential activity. That this should also, again, be reducing a certain kind of theoretical effort (that of the imagination, Leibniz would have said) to a minimum, may sound paradoxical indeed, in the light of the frustration one feels by opening, say, the *Grundlagen der Ontologie* or the *Grundzüge*, where the directives are laid down, preceded by a terrifying list of carefully specified terminological explanations of the terms occurring in the directives, plus a vocabulary to understand the terms occurring in the explanations. But the end of all this is clear: the rules to pass from axioms to theorems, like the axioms, rest upon nothing other than the meanings of the terms used in them, that is, they are (in this sense) self-evident.

[21] This frequently missed point is stressed by Sundholm, cf. (2003: 110, not directly about Leśniewski).

So much for the necessity of the axioms. How about that of the theorems? One way to make sense of the necessity requirement is, traditionally, to link it to aprioricity.[22] There is not much in Leśniewski's *œuvre* that relates to this notion explicitly. But, at one early point, Leśniewski calls "*a priori*" propositions

> whose truth can be demonstrated by means of linguistic conventions alone or propositions which follow from these conventions
>
> (Leśniewski 1912: 219–20; English translation 1991: 39–40, amended).

Once again, the passage is from an early paper, and in this case we must keep this in mind. But the passage is, in itself, interesting. At first, it sounds to our modern ears as if Leśniewski here had in mind not aprioricity, but rather the much later notion of *analyticity* as truth in virtue of meaning-rules alone. Two things are worth mentioning in this connexion.

First, the "linguistic conventions" Leśniewski has in mind here are not solely what *we* would call meaning-rules (like the one governing the behaviour of negation, for instance), but heavy-duty logical assumptions encoding, for example, a version of the principle of contradiction as well as a semantic assumption yielding a fairly strong version of the correspondence theory of truth.[23]

Secondly, with this in mind, if for "linguistic conventions" we substitute "general laws", Leśniewski's view appear to be akin to Frege's, as expressed in the following passage:

> For truth to be a posteriori, it must be impossible to construct a proof of it without including an appeal to facts, i.e., to truths which cannot be proved and are not general, since they contain assertions about particular objects. But if, on the contrary, its proof can be derived exclusively from general laws, which neither need nor admit of proof (*weder fähig noch bedürftig sind*), then the truth is a priori.
>
> (Frege 1884: 4, English translation).[24]

---

[22] For instance, in Kant, cf. De Jong (2001: 331).

[23] Leśniewski (1912: 215, 216–7; 1991: 34–5, 36). This is an example of how Leśniewski looks at his 'conventions': "§15. Science is, then, a system of linguistic symbols. Creating and understanding linguistic symbols require [...] certain principles of creation for symbols and keys to decipher symbols [ ]. Linguistic conventions are, therefore, the necessary condition of any scientific creativity and the indispensable key to understanding science. [...]". Leśniewski (1912: 216; English translation 1991: 35, amended). Three things are worth noting: (1) one possible way to look at Leśniewski's conventions in the light of the Model, given that in his early writings Leśniewski does not see logic but rather ontology as the most fundamental Aristotelian Science, is to see them as some kind of ultimate principles; (2) the passage is relevant to understand Leśniewski's inclination from the very beginning towards a semantical take on the notion of domain in the sense explained in Section 4; (3) of the correspondence theory of truth no trace will be left in his logical systems: this is one of reasons why it is important to keep in mind that this is an early remark.

[24] The "general laws" Frege has in mind are the fundamental propositions of the Model – note, *en passant*, that with this the 'unclarity' Woleński (2004a: 813) attributes to Frege on this point seems to vanish. The famous expression "weder fähig noch bedürftig sind" applied to principles is already in Eduard Zeller's *Die Philosophie der Griechen in ihrer geschichtlichen Entwicklung dargestellt* II, Part II *Abteilung Aristoteles und die alten Peripatetiker*, 2 Lfg., Tübingen, 1860–1862, cf. W. Detel's *Aristoteles – Analytica posteriora*, übersetz. und erläut. von Wolfgang Detel, Berlin, Akademie Verlag, 1993: 267–8.

So, Leśniewski seems to follow a perspective similar to Frege in linking the *a priori* with justification from (non-factual) general laws. But I do not wish to suggest that he got this directly from Frege. On the contrary, Leśniewski in 1911–2 was not yet conversant with Frege's work.

Note, *en passant*, that in the few places in which he does mention it, Leśniewski has a deviant view of the analytic/synthetic distinction and its interrelation with the *a priori/a posteriori*, at least with respect to Kant. Analytic propositions are for Leśniewski those (in "canonical form", that is in subject-predicate form, with the positive form of the copula "is") that "contain no predicates which connote properties that are not connoted by the subject", synthetic those that "contain predicates which connote also such properties that are not connoted by the subject" (Leśniewski 1911: 330–1; English translation 1991: 2–3). It follows ultimately that for him true a priori propositions can be only *synthetic* propositions and propositions which are based solely on synthetic propositions (Leśniewski 1912: 222; English translation 1991: 42):

The conclusion follows that all analytic propositions true *a priori* are based on the truth of synthetic propositions. This conclusion can be of some value for the theory of science. (Leśniewski 1912: 222; English translation 1991: 42, amended)

Interestingly, Leśniewski explicitly connects the (a priori) ordering of truths of a science with the analytic/synthetic distinction. This is very much in line with the way in which the Aristotelian Model of science. But as far as I know, Leśniewski did not develop his position on the *a priori/a posteriori* and the analytic/synthetic distinctions any further in his later writings. Although Leśniewski seems to follow Frege in the early passages on the *a priori*, he came into contact with Frege's work only later, after the First World War and probably not before 1918. Was it someone else who inspired Leśniewski then? Despite the Bolzanian ring of the early passages quoted in the last section, we have too little evidence that his direct source was Bolzano. But the latter could have been his indirect source through Husserl. And indeed Husserl was, most likely, the most important source for the Model Leśniewski had at his disposal at that time, in particular Husserl's *Prolegomena* to the *Logical Investigations* (1900/01):

> §63. Scientific knowledge is, as such, *grounded knowledge* [...]. The grounding of general laws necessarily leads to certain laws which in their essence, i. e. intrinsically, and not merely subjectively or anthropologically, cannot be further grounded. These are called basic laws (*Grundgesetzen*)
>
> (Eng. Tr.: 228, amended).

This passage is strikingly similar to the above quoted one by Frege. Husserl's decisive step towards a conception of logic as a science of science, or, to use Bolzano's expression, as a theory of science, had an important and long-lasting influence on Leśniewski's thought.[25] So, Leśniewski follows a Frege-like, or more

[25] Leśniewski read Husserl before he moved to Poland to study with Twardowski. I presume he got into contact with Husserl's thought in 1910 during his studies in Munich where Alexander Pfänder based his exercise sessions in logic on the *Logical Investigations*. This means that Leśniewski's adherence to the Model has nothing to do with the historiographical *topos* that is normally brought

plausibly from the historical point of view, a Husserl-like perspective in linking the *a priori* to justification from (non-factual) general laws. Therefore, even if in the later Leśniewski aprioricity does not seem to play an explicit role, if aprioricity is interpreted as freedom from empirical assumptions and as following from most general laws, then it can well be said that all three of Leśniewski's systems are *a priori*, thanks to their generality and ontological neutrality.

6.6 Conclusion

I hope to have shown convincingly that the relationship of Leśniewski's systems to the Model is one of deep closeness. It is easier to show in what sense Leśniewski's systems obey conditions II, III and IV, and more problematic to show in what sense they follow I and V, but the problems one faces in applying I and V prove intrinsic to the historical development of those conditions themselves. As for I, this is the condition the rationale of which has changed the most in the course of history, one of the most profound changes being due to the evolution of the status of logic as a science, while V is the condition that philosophers have found the most problematic in itself.

Leśniewski, following Frege, built his system as a continuation of Leibniz's dream of *characteristica universalis*, an all-encompassing science of science, where the science is an Aristotelian science in the sense of the Model. Showing how the notions of *calculus ratiocinator* and *lingua characteristica* exactly connect to the Model as realized in Leśniewski's systems would offer a particularly interesting perspective on both Leśniewski's and Frege's conceptions of science. This, as mentioned, I shall leave for another occasion.[26] At any rate, on the basis of what precedes this much can be said: according Leśniewski's take on the Model, all you

---

in to explain the presence of 'Aristotelian' elements in any exponent that has even loose links with the Austrian tradition, that is, the appeal by default to the generic Brentano factor.

[26] The clash that van Heijenoort's framework aims at capturing centres on whether logical *calculi* are or are not, should or should not also be, *linguae characteristicae*. Frege saw his *Begriffsschrift* as being both: indeed in light of the Model it becomes possible to see that the two notions are not supposed to exist independently of each other. This is clear in the picture of Leibnizian logic given by Trendelenburg, Květ and Exner in the three German studies devoted to Leibniz before Frege's *Begriffsschrift*. In Trendelenburg's "Ueber Leibnizens Entwurf einer allgemeinen Characteristik" (1867), we read: "Every proof presupposes definitions. As a matter of fact, the ultimate principles are definitions and statements of identity, i.e. judgements which are proved analytically from the identity of concepts. The important thing is that we form suitable definitions by using sign formulas [of *lingua characteristica*] so that they can form a foundation for the inferring calculus, *calculus ratiocinator*." (Quoted from Korte (forthcoming: 5) to whom I owe my information on Trendelenburg, Květ and Exner) Trendelenburg presents a *calculus ratiocinator* as a proof device operating on ultimate principles formulated in a *lingua characteristica*. Trendelenburg's mention of proof, definitions, concepts, ultimate principles and their arrangement presupposes precisely the Model. The relationship between *calculus* and *lingua* is described in such a way that a connection is immediately established between the first notion and the Model Postulates II and III, and between the second notion and Postulates I, IV and V. Leśniewski's enterprise is in line with this.

need to understand to create mechanically (Leibniz would say *blindly*) a indefinitely large number of truths, is the following: a single concept of the lowest grammatical degree placed in the initial segment of a single axiom, the latter being one single proposition with the form of an equivalence, the shortest possible, with the fewest amount of variables possible. The most important requirement for such a collection of propositions to be a *science* based on a *language* and not just on a *calculus*, however, is that the primitives be *meaningful* and the axioms be *true*.

Acknowledgments Work on this paper has been funded by the Netherlands Organisation for Scientific Research (Project 275-80-001). Thanks to Wim de Jong, Tapio Korte, Katarzyna Kijania-Placek, Sandra Lapointe, Marije Martijn, Peter Simons, and Göran Sundholm for helpful comments on earlier versions of this paper. See also note 5.

References

Barnes, Jonathan (1999) *Aristotle – Posterior Analytics*, Translated and with a Commentary by Jonathan Barnes, Oxford, Clarendon.

Beaney, Michael (2003) 'Analysis', *The Stanford Encyclopedia of Philosophy (Summer 2003 Edition)*, Edward N. Zalta (ed.), URL = <http://plato.stanford.edu/archives/sum2003/entries/analysis/>.

Beth, Evert W. (1959) *The Foundations of Mathematics – A Study in the Philosophy of Science*, Amsterdam, North-Holland.

Betti, Arianna (2004) 'Leśniewski's Early Liar, Tarski and Natural Language', *Annals of Pure and Applied Logic*, 127, 267–287.

Betti, Arianna (forthcoming) 'Leśniewski's *Characteristica Universalis* and the Classical Model of Science' *Synthese*.

De Jong, Willem R. (1996) 'Gottlob Frege and the Analytic-Synthetic distinction within the framework of the Aristotelian Model of Science' *Kant-Studien* 87, 290–324.

De Jong, Willem R. (2001) 'Bernard Bolzano, Analyticity and the Aristotelian Model of Science' *Kant-Studien*, 328–349.

De Jong, Willem R. and Arianna Betti (forthcoming) 'The Classical Model of Science: a Millennia-Old Model of Scientific Rationality', *Synthese*.

De Pierris, Graciela (1988) 'Frege and Kant on a priori knowledge', *Synthese* 77, 285–319.

Frege, Gottlob (1884) *Die Grundlagen der Arithmetik – Eine logisch mathematische Untersuchung über den Begriff der Zahl*, Breslau, Breslau, W. Koebner. Reprinted together with the English Translation (J. L. Austin) *The Foundation of Arithmetic*, Oxford, Blackwell,1950.

van Heijenoort, Jean (1967) 'Logic as Language and Logic as Calculus', *Synthese* 17, 324–30.

Husserl, Edmund (1900/01) *Logische Untersuchungen*, Max Niemeyer: Halle a.d.S., 1900–1901; = *Husserliana* XVIII, XIX/1, XIX/2, The Hague, Nijhoff. (Eng. tr. by J. N. Findlay, London/Henley, Routledge, 1970.)

Kambartel, Friedrich (1975) 'Frege und die axiomatische Methode. Zur Kritik mathematikhistorischer Legitimationsversuche der formalistischen Ideologie', in: M. Schirn (ed.), *Studien zu Frege*, Band I, Stuttgart, Frommann-Holzboog, 1976, 215–228.

Korte, Tapio (forthcoming) '*Begriffsschrift* as a *lingua characteristica* and the Classical Model of Science', *Synthese*, forthcoming.

Küng, Guido (1981), 'Leśniewski's Systems', in W. Marciszewski (ed.), *Dictionary of Logic*, The Hague, Nijhoff, 168–177.

LeBlanc, A. O. V. (1983) *A Study of the Axiomatic Foundations of Mereology*, MA thesis, Faculty of Arts, University of Manchester.

LeBlanc, A. O. V. (1991) *Leśniewski's Computative Protothetic*, PhD Thesis, Faculty of Arts, University of Manchester,

Lejewski, Czesław (1954) 'Logic and Existence', in J. T. J. Srzednicki & Vito F. Rickey (eds.) (1984) *Leśniewski's Systems – Ontology and Mereology*, The Hague, Nijhoff, 45–58

Lejewski, Czesław (1982) 'Ontology: What Next', in: W. Leinfellner, E. Kraemer and J. Schank (eds.), *Language and Ontology. Proceedings of the 6th International Wittgenstein Symposium*, Vienna: Hölder-Pichler-Tempsky, 173–185

Leśniewski, Stanisław (1911) 'Przyczynek do analizy zdań egzystencjalnych' *Przegląd Filozoficzny* XIV, 329–45. Eng. tr. in Leśniewski (1991), 1–19.

Leśniewski, Stanisław (1912) 'Próba dowodu ontologicznej zasady sprzeczności', *Przegląd Filozoficzny* XV, 202–26; Engl. transl. in Leśniewski (1991), 20–46.

Leśniewski, Stanisław (1916) *Podstawy ogólnej teorji mnogości. I* (Część. Ingredyens. Mnogość. Klasa. Element. Podmnogość. Niektóre ciekawe rodzaje klas), *Prace Polskiego Koła Naukowego w Moskwie*, Sekcja matematyczno-przyrodnicza. Engl. trans. in Leśniewski (1991), 129–73.

Leśniewski, Stanisław (1927/31), *O Podstawach Matematyki*, Rozdz. I-III, *Przegląd Filozoficzny* XXX, 164–206, 1927; Rozdz. IV, XXXI, 1928, 261–91, Rozdz.V–IX, XXXII, 1929, 60–101, Rozdz. VI–IX, XXXIII, 1930, 77–105, Rozdz. X–XI, XXXIV, 1931, 142–170. Eng. Tr. in Leśniewski (1991), 181–382.

Leśniewski, Stanisław (1929) "Grundzüge eines neuen Systems der Grundlagen der Mathematik", *Fundamenta Mathematicae* XIV, 14. (Eng. Tr. in (1991, 422.)

Leśniewski, Stanisław (1939) "Einleitende Bemerkungen zur Fortsetzung meiner Mitteilung u. d. T. 'Grundzüge eines neuen Systems der Grundlagen der Mathematik"', *Collectanea Logica*, 1, 1–60; Eng. Tr. in Leśniewski (1991), 649–710.

Leśniewski, Stanisław (1991) *Stanisław Leśniewski.Collected Works*, 2 vols., ed. by S. J. Surma, J. T. Srzednicki, D. I. Barnett and V. F. Rickey, Dordrecht, Kluwer.

Lindenbaum, Adolf (1936) 'Sur la simplicité formelle des notions' *Actes du congres international de philosophie scientifique*, VII Logique, Paris, Hermmann, 29–38.

Luschei, Eugene C. (1962) *The Logical Systems of Leśniewski*, Amsterdam, North-Holland.

Miéville, Denis (1984) *Un développement des systèmes logiques de Stanislaw Lesniewski, Protothétique – Ontologie – Méréologie*, Bern/Frankfurt/New York, Peter Lang.

Mignucci, Mario (1975) *L'argomentazione dimostrativa in Aristotele – Commento agli Analitici Secondi* I, Padova, Antenore.

Philipse, Herman (1983) *De fundering van de logica in Husserls 'Logische Untersuchungen'*, PhD diss., Leiden.

Philipse, Herman (2004) 'Edmund Husserl and the History of Classical Foundationalism', in: R. Feist (ed.), *Husserl and the Sciences: Selected Perspectives*, Ottawa, University of Ottawa Press, 11–40

Scholz, Heinrich (1975) 'The Ancient Axiomatic Theory' translated in J. Barnes, M. Schofield, and R. Sorabji (eds.) *Articles on Aristotle*. Volume 1: Science, London: Duckworth, 1975, PAGES (= "Die Axiomatik der Alten"*Blätter für Deutsche Philosophie IV*, 1930, 259–278.)

Sobociński, Bolesław (1955/6) 'On well-constructed Axiom Systems' *VI Rocznik Polskiego Towarzystwa Naukowego na Obczyźnie*, 54–70.

Sundholm, Göran (2003) 'Tarski and Leśniewski on languages with meaning versus languages without use – A 60th Birthday Provocation for Jan Woleński', in: J. Hintikka, T. Czarnecki, K. Kijania-Placek, T. Placek, A. Rojszczak† (eds.) *Philosophy and Logic – In Search of the Polish Tradition*, Essays in honour of Jan Woleński on the Occasion of his 60th Birthday, Dordrecht, Kluwer, 109–128.

Tarski, Alfred (1923). 'O wyrazie pierwotnym logistyki' *Przegląd Filozoficzny* 26: 68–89. (Eng. trans. in *Logic, Semantics, Metamathematics – Papers from 1923 to 1938*. Oxford, Clarendon Press: 1–23.)

Woleński, Jan (2004a) 'Analytic vs, Synthetic and Apriori vs. Aposteriori' in I. Niiniluoto, M. Sintonen and J. Woleński, Handbook of Epistemology, Dordrecht, Kluwer, 781–839.

Woleński, Jan (2004b) 'The Reception of Frege in Poland' *History and Philosophy of Logic* 25, 37–51.

Chapter 7
Leśniewski, Negation, and the Art of Logical Subtlety

Denis Miéville

> *It is a curious paradox, puzzling to the symbolic mind, that definitions, theoretically, are nothing but statements of symbolic abbreviations, irrelevant to the reasoning and inserted only on for practical convenience, while yet, in the development of a subject, they always require a very large amount of thought, and often embody some of the greatest achievements of analysis*
>
> (Russell 1903: 63).

7.1 Introduction

Leśniewski essentially developed three theories: Protothetic, Ontology, and Mereology. Since his death in 1939, none of the efforts to reawaken interest in Leśniewski have had much success. In spite of his successive burials, I am among those who persevere in thinking that Leśniewski's systems present more than a merely historical interest. The richness of Leśniewski's alternative lies in his approach to truth and falsity, the idea of predicative levels and his conception of logic as something which, so to say, freely "expands". Leśniewski's systems can be called to task when it comes to the study of formal languages, the development of higher order logics, definitional procedures, the search for extreme metalinguistic rigor and the quest for an ontologically neutral language. In this paper, I focus on the following three issues. First, I consider the question of the number of operators a formal system must or can possess. Secondly, I argue that those unsatisfied by the conceptual paucity of classical logics – that is to say, systems that were initially developed specifically as tools for the foundations of arithmetic – should envision a new way of developing formal systems, and that Leśniewski's work has, in this respect, valuable heuristic potential. Finally, after presenting the main lines of Leśniewski's project, I will illustrate the value of his approach. Although Leśniewski was a marginal figure, as far as both his work and his character are concerned, he occupies a central position in the history of Polish philosophical thought. The fact that he studied and collaborated, namely with Twardowski, Adjukiewicz, Łukasiewicz, and Tarski is not insignificant

D. Miéville (✉)
Université de Neuchâtel (Suisse), Neuchatel, Switzerland
e-mail: denis.mieville@unine.ch

S. Lapointe et al. (eds.), *The Golden Age of Polish Philosophy*, Logic, Epistemology, and the Unity of Science 16, DOI 10.1007/978-90-481-2401-5_8,

in this respect (see for instance the papers by Łukasiewicz, Patterson, Woleński, this volume).

7.2 Logical Operators

Are there logical functors other than those suggested by classical logics? In particular, are there functors other than those we find in propositional and predicate logic that could be attractive to logicians who are concerned with the operations involved in rational thought processes? The answer is, of course, "Yes". The principle of obversion, for instance, which was first proposed by Boole – but who received its name from Alexander Bain – is just one such operation:

> A universal-affirmative proposition is convertible into a universal-negative, and vice versa by negation of the predicate. (. . .) A particular-affirmative proposition is convertible into a particular-negative, and vice versa by negation of the predicate.
>
> (Boole 1965: 29; first edition 1847)

Of course, this conception of negation sounds like trouble as soon as we relate it to first order predicate logic, which requires that all predicates be universally applicable. As Corcoran explains:

> The predicate prime is true of two, false of four and not applicable to pi. This means that such predicate has a range of applicability within which it holds true or false and outside of which it does not hold at all. Thus a sentence can fail to be true without being false and it can fail to be false without being true... Yet the following is logically true in standard logic: $(\forall X)(PX \vee \sim PX)$. This reflects the fact that standard semantics presupposes universal ranges of applicability for all predicates.
>
> (Corcoran 1973: 43)

Thus, within the perspective of classical logic, *not belonging to the extension of a property* necessarily entails *belonging to the complementary extension*. This is a weakness inasmuch as it is not possible to express, in the syntax of this logic, that every object is associated, within a universe of reference, with a particular domain of determinations which are its own, and that some other determinations in no way concern them. There must be some interest in having at our disposal a logical language able to express in a precise way the fact that for any given object, say, for example, the number 5 there is a distinction to be made between the properties that belong to the domain of that object, for example, *to be an even number / to be a un-even (odd) number* and properties that do not concern this object such as, in this case, *to be moral / to be immoral*. In most semantics, the following two expressions are considered to be true:

Five is not an even number
Five is not moral

If we accept that *negation is the declaration that one thing is separated from another thing*, as Aristotle said (*De Interpretatione*, 6. 17a: 26–27), these two propositions however differ as to the nature of this separation. Consider a logic such that while it takes into account the principle of excluded middle, it also contains a principle which can be formulated in the following manner:

For all property *p* and for all object x, EITHER (x is associated either with the property *p* or with the dual property of *p*) OR (x is neither associated with the property *p* nor with the dual property of *p*)

To meet this condition, we need to make use of a new negation which I will call "nominal negation". This negation applies to names and violates the law of obversion. Such a negation epitomizes what Aristotle meant when he wrote:

> If everything is equal or is not equal, everything is not equal or unequal, apart from the subject capable of receiving the inequality.
>
> (Aristotle, Cat. 1055b: 10)

Accordingly, we need a new logical connector, absent from classical logics. And if we inquire further, we see that other logical operators may be necessary to describe certain other deductive articulations. My investigation into the domain of negation operators can be transposed to other types of logical operators such as, for example, conjunction and disjunction. In fact, the fundamental question is the following: how many operators could we define if we were to retain only the principles of bivalence and non-contradiction, and if our aim was precisely to generate an indefinite number of new functors and new relations. The answer I would like to give is that there would be, in principle, as many operators as it is possible to define on the basis of the two basic categories of *proposition* – here the term "sentence" should be seen to be indifferently synonymous – and *name*, S and N, respectively. Thus, in this non-restrictive and indeed liberal perspective we aim at a logic that will enable us to identify as directly as possible (i) every possible logical functor belonging to any syntactico-semantic category arising from a combination of categories N and S, (ii) the derivation of every thesis which might concern them. How do we achieve this programme? With what type of formal system can we realize this project? On what axiomatic basis should we establish it? Leśniewski's programme provides an answer to all these questions. In what follows, we are following Leśniewski.

7.3 A Possible Answer

Standard formal languages are unable to capture every syntactico-semantic category conceivable on the basis of the basic categories S and N. The classical approach to formal systems gives the set of well-formed expressions inductively, on the basis of a finite list of primitive terms and recursive formation rules; it supposes a limited set of logical constants. But a Leśniewskian-type formal language cannot be syntactically determined from the start. The language must allow its syntax to be "developed" or "expanded". We do not initially have at our disposal an exhaustive list of symbols corresponding to the set of all possible logical constants – nor do we actually have a set of variables. The idea behind a Leśniewskian-type language is to develop the syntax without beforehand giving a definitive list of symbols. We must warrant that the definition of logical concepts be developed as the system progresses and insure that they'd be compatible with the principles of non-contradiction and bivalence. To grasp better the phenomenal richness of this potential categorial expansion, let consider the Leśniewskian-style grammar below:

(a) S and N are syntactico-semantic categories.
(b) If $C_1, \ldots, C_n$, and also C_k are syntactico-semantic categories, then $(C_k/\ C_1 \ldots C_n)$ is a further syntactico-semantic category of expressions which, taking expressions of the categories C_1 to C_n as arguments, form expressions of category C_k. So, for instance, the category (S/N) is the category of expressions that, given an N as an argument, form an expression of category S—the category, that is, of one-place predicates.
(c) Nothing is a syntactico-semantic category except by what precedes.

Such a categorial extension would include expressions of the following form: (S/S), (N/N), (N/N)/N, (N/N)/ (N/N), (N/S), (N/S)/ (N/S), (S/N), (S/(S/N)), (S/((N/N)/N), (S/NS)/(S/N), for instance.

Leśniewski proposed an axiomatic basis which is extremely modest in terms of primitive terms. In fact, it contains only two primitive terms. The first is the one associated with the biconditional operator of category S/SS. The second is the one underlying the logical term for a singular proposition "A ε b" and which means that the object designated by A is among those designated by the name b, or is the object designated by b itself.[1]

A Leśniewskian-type grammar contains, in the first instance, only the finite number of syntactic characters that convey the fundamental concepts. Every new symbol, every new expression is determined not as a function of an *a priori* list of predetermined types of symbols. Rather a new expression is introduced by the logician who expands the system on the basis of the fundamental principles of definition. The process is both constructive (it follows determinate definitional principles) and creative.

Leśniewski's chose the biconditional as the sole primitive propositional term. This is easy to justify. First, every explicit definition states that a *definiendum* A and a *definiens* B are in an equivalence relation.

$$A \leftrightarrow B$$

In a logical system, the validation of such equivalences rest on the availability, as a theorem, of a biconditional operating on A and B:

$$A \leftrightarrow B, \text{ if and only if } \vdash A \equiv B$$

[1] Note that contrary to what is the case in Husserl, who was read by Leśniewski and whose views on categorial grammar he influenced, the use of uppercase and lowercase letters in the place of subject and predicate is not meant to signify a categorial difference: both stand for names.(Cf. *Logical Investigations*, fourth *Investigation*, Halle/Salle, Niemeyer, 1901) Husserl's – and Leśniewski's – notation can be traced back to Bolzano (*Wissenschaftslehre* §60, Sulzbach, Seidel, 1837)

The second point of justification is a result we owe to Tarski. In his doctoral thesis completed under the direction of Leśniewski in 1923, Tarski demonstrates the following: within the context of a logical system containing the biconditional as the unique functor and which allows quantifiers to bind variables of propositional categories, it is possible to define both conjunction and negation. By turning definitions into theses of the system – in the form of bi-conditional sentences – we avoid their having the status of pure abbreviation that they have in Russell's standard kind of definitions. Since they don't have that status, definitions can be used to introduce creative logical concepts in the development of the logical theory.

In a system of the type of Leśniewski's, the copula "ε" has little in common with the epsilon "∈" of set theory. "ε" is precisely meant to deal with extensions without appealing to set-theoretical notions since the terms it puts in relation designate objects (not sets). On the basis of these foundational ideas, Leśniewski proposes a formal method for constructing logical theories on the basis of a finite list of primitive symbols (with primitive meanings). By means of appropriate inference rules – in particular, the rules for definitions – the system allows for any functor of any syntactico-semantic category conceived in terms of the primitive categories – S and N – and it does so consistently. This conception of formal language is extremely liberal as far as syntactic choices are concerned. This liberality follows from the fact that categorial determination does not lie in the form of signs, but in the "context" in which they appear. For example, the use of the sign "≡" for the biconditional Leśniewski has in mind is not sufficient to determine fully the intended category, in this case S/SS. The sign carries its specific signification, without equivocation, only once it appears in a given context. In Leśniewski's notation, a context is signified by a determinate type of bracket pair. For instance, when a sign or inscription immediately precedes the schema "(. . . . . .)", it belongs to the category S/SS. In Leśniewski, the context in which ε appear is {. . . . . .} and expressions of this type belong category S/NN. The expansion of a theory is the result of judiciously choosing new contexts. Any new context may be introduced, provided it does not imply a contradiction or create an ambiguity. This expanding grammar (progressively) yields all possible functors. It also allows for polysemy and synonymy. Thus, if I set up definitions for two negations of different categories, I can use the same symbol without risk of confusion, ambiguity, or contradiction. I just need to choose, for each of them, different contexts, for instance "∼ (. . .)" for the nominal negation (N/N) and "∼< . . . >", for propositional negation (S/S). The axiomatic basis constitutes the only syntactic set initially stated in the system. This basis is of course finite: no list of symbols is postulated outside of it.

7.4 A Final Example

Let us define, on the basis of Leśniewski's system, a new negation that will eliminate the law of obversion while conserving the law of excluded middle. This negation is to act on unary predicates and is such as to epitomize the difference between objects which, in a determinate semantic system, can be associated with some notion and those which are excluded by this notion. The extension of a predicate is determined by the duality between objects which have the predicate and objects which are “concerned” by the predicate, but do not have it. This extension is further determined by the exclusion of all objects not concerned by the predicate in question. Thus, the predicate “even number” in a semantic model which would contain both elementary arithmetic and geometrical entities, would be associated with the extension: “, 2, 4, 6, 8, 10, etc.”, on the one hand and its complement, “1, 3, 5, 7, 9, etc.” on the other. But the complement of this predicate would not include, say, straight lines, figures and solids and the predicate itself would therefore not concern the latter.

Following Leśniewski, let us now consider the following definition for a non propositional negation:

$$(\forall Ab)(A \text{ is } \sim< b > \equiv ((A \text{ is } A) \wedge \sim (A \text{ is } b)))$$

$$\lfloor Ab \rfloor \lceil \equiv (\varepsilon\{A \sim< b >\} \wedge (\varepsilon\{AA\} \sim (\varepsilon\{Ab\})))\rceil$$

The second expression is a reformulation of the first in a Leśniewskian-type contextual (bracket) notation and the inscription $\lfloor Ab \rfloor$ is used for quantification. This definition, which constitutes a thesis of the system, includes two negation operators: “$\sim$” before the context “$< \ldots >$”, which maps a formative operator of the category of names to a nominal argument, (N/N); and “$\sim$” which maps a formative operators of the category of proposition to a propositional argument, represented by the context “(. . .)”. There is an air of banality to this definition, for while it defines a new negation, it seems not to bring any new meaning into play. However, were there to be no new meaning, we would not be able to achieve the results we want, namely, rejecting the law of obversion (which provides an elucidation of the law of “appropriateness”) while respecting the fundamental principles of traditional logic. But let us consider what follows.

Principle of non-contradiction

1. $(\forall Ab)(\sim (A \text{ is } \sim< b > \wedge \sim (A \text{ is } \sim< b >))$
   $\lfloor Ab \rfloor \lceil \sim (\wedge(\varepsilon\{A \sim< b >\} \sim (\varepsilon\{A \sim< b >\})))\rceil$
2. $(\forall Ab)(\sim (A \text{ is } b \wedge \sim (A \text{ is } b))$
   $\lfloor Ab \rfloor \lceil \sim (\wedge(\varepsilon\{Ab\} \sim (\varepsilon\{Ab\})))\rceil$

Principle of singular conditional

1. $(\forall Ab)(A \text{ is } b \supset \sim (A \text{ is } \sim< b >))$
   $\lfloor Ab \rfloor \lceil \supset (\varepsilon\{Ab\} \sim (\varepsilon\{A \sim< b >\}))\rceil$
2. $(\forall Ab)(A \text{ is } \sim< b > \supset \sim (A \text{ is } b))$
   $\lfloor Ab \rfloor \lceil \supset (\varepsilon\{A \sim< b >\} \sim (\varepsilon\{A < b >\}))\rceil$

Principle of contrariety

$(\forall Ab)(\sim (A \text{ is } b \wedge A \text{ is } \sim< b >)$
$\lfloor Ab \rfloor \lceil \sim (\wedge(\varepsilon\{Ab\}\varepsilon\{A \sim< b >\}))\rceil$

Principle of particular conditional

1. $$\sim (\forall A)(\forall b)(\sim (A \text{ is } \sim< b >) \supset A \text{ is } b))$$
$$\sim \lfloor Ab \rfloor \lceil \supset (\sim (\varepsilon\{A \sim< b >\})\varepsilon\{A\ b\})\rceil$$

2. $$\sim (\forall A)(\forall b)(\sim (A \text{ is } b) \supset A \text{ is } \sim< b >))$$
$$\sim \lfloor Ab \rfloor \lceil \supset (\sim (\varepsilon\{A\ b\})\varepsilon\{A \sim< b >\})\rceil$$

Principle of non-compatibility

$$\sim (\forall A)(\forall b)(A \text{ is } b \vee A \text{ is } \sim< b >)$$
$$\sim \lfloor Ab \rfloor \lceil \vee(\varepsilon\{A\ b\}\varepsilon\{A \sim< b >\})\rceil$$

Principle of excluded middle revisited

$$(\forall Ab)((A \text{ is } b \text{ w } A \text{ is } \sim< b >)\vee \sim ((A \text{ is } b \text{ w } A \text{ is } \sim< b >)))$$
$$\lfloor Ab \rfloor \lceil \vee(\text{w}(\varepsilon\{A\ b\}\varepsilon\{A \sim< b >\}) \sim (\text{w}(\varepsilon\{A\ b\}\varepsilon\{A \sim< b >\})))\rceil$$

These results, all of which are theorems, allow us to violate the law of obversion which

> [c]auses the disappearance of the linking of two terms in favor of complementation, which doesn't happen without some shifting trick: purely formal contradiction is no longer associated with semantic contrariety. The distribution of things into classes henceforth overtakes the articulation of thoughts.
>
> (Frey 1987: 60)

By making use of the constructive defining power of Leśniewski, we can break through the constraints of bipolar sets in favor of a principle of "appropriateness" that restores what some consider to be an indispensable aspect of rational thought. Nothing keeps us from pursuing this analysis while projecting it on other syntactico-semantic categories, which a Leśniewskian methodology easily permits (Miéville 1991).

7.5 Conclusion

I have attempted to sketch Leśniewski's conception of a formal system. It consists in the inscription of a finite number of primitive symbols and contexts. By virtue of determinate inferential directives, we can progressively define any functor of any syntactico-semantic category resulting from a combination of the two basic categories of propositions and names. This kind of system is doubly open: the syntactic level, its expansion is both quasi limitless and intelligently restrictive; moreover every expansion is associated with a semantic development. This controlled liberality is thus an extraordinary tool for anyone who ventures into the clarification of the possible moves put to work in the exercise of rational thought. The metalanguage of these systems has been formalized by Leśniewski and axiomatized by Rickey (1972 and 1973).

References

Bain, Alexandre (1875), *Logique déductive et logique inductive*, vol. 1–2, Paris, Gernier Boillère.
Boole, George (1847), *The Mathematical Analysis of Logic*, Oxford, Basil Blackwell, 1965.
Castaneda, Hector-Neri (1988), 'Negations, imperative, colors, indexical properties non-existence, and Russell's', in D.F. Austin (ed.), *Philosophical Analysis, A Defense by Example*, Dordrecht, Kluwer, 169–205.
Corcoran, John (1973), 'Gaps between logical theory and mathematical practice', in M. Bunge (ed.), *The Methodological Unity of Science*, Dordrecht, Reidel, 23–50.
Frey, Louis 1(988), 'De la négation à l'affirmation en logique aristotélicienne', *Travaux du Centre de Recherches Sémiologiques*, n 56, 121–138.
Gessler, Nadine (2005), *Introduction à l'œuvre de S. Leśniewski III, La Méréologie*, Université de Neuchâtel, Travaux de Logique.
Leśniewski, Stanisław (1992), *Collected Works 1–2*, S.J. Surma, J.T. Srzednicki, D.I. Barnett (eds.), Warsaw/Dordrecht, PWN/Kluwer.
Miéville, Denis (1984), *Un développement des systèmes logiques de S. Leśniewski. Protothétique. Ontologie. Méréologie*, Berne, Lang.
Miéville, Denis (1991), *La négation, une étude logique*, Université de Neuchâtel, Travaux de Logique 6.
Miéville, Denis (2001), *Introduction à l'œuvre de S. Leśniewski I, La protothétique*, Université de Neuchâtel, Travaux de Logique.
Miéville, Denis (2004), *Introduction à l'œuvre de S. Leśniewski II, L'ontologie*, Université de Neuchâtel, Travaux de Logique.
Miéville, Denis (2006), 'Leśniewski, une manière originale de penser le métalangage', in R. Pouivet & M. Rebuschi (eds.), *La philosophie en Pologne 1918–1939*, Paris, Vrin, 223–246.
Peeters, Marc (2006), *Introduction à l'œuvre de S. Leśniewski IV, L'œuvre de jeunesse*, Université de Neuchâtel, Travaux de Logique.
Rickey, Frederick (1972), 'Axiomatic inscriptional syntax, part I, General syntax', *Notre Dame Journal of Formal Logic* 13, 1–33.
Rickey, Frederick (1973), 'Axiomatic inscriptional syntax, part, II, Syntax of protothetic', *Notre Dame Journal of Formal Logic* 14, 1–52.
Russell, Bertrand (1903), *The Principles of Mathematics*, London, Allen & Unwin.
Simons, Peter (1987), *Parts. A Study in Ontology*, Oxford, Clarendon Press.

Chapter 8
Philosophy of Mathematics in the Lvov-Warsaw School

Roman Murawski

8.1 Philosophers and Mathematicians in Lvov and Warsaw

The aim of the paper is to present and discuss the main views and tendencies in the philosophy of mathematics of the members of the Lvov-Warsaw Philosophical School and of the Polish Mathematical School, and to assess the influence of these views on logic and mathematics. One should start by stressing, on the one hand, the fact that logicians and mathematicians in Warsaw in the interwar period closely collaborated and mutually influenced each other and, on the other, that Warsaw mathematicians were interested in philosophical problems concerning mathematics and, vice versa, philosophers and logicians were open to mathematics and its philosophical problems. The collaboration and mutual understanding between philosophers and logicians and mathematicians in Warsaw was strong. Indeed, Jan Łukasiewicz and Stanisław Leśniewski (who were philosophers!) had chairs at the Mathematical Faculty of Warsaw University (Łukasiewicz held the chair in logic and Leśniewski the chair in the philosophy of mathematics). What's more, the two leading figures of the Warsaw Mathematical School, Wacław Sierpiński and Zygmunt Janiszewski chose philosophy of mathematics as the subject of their habilitation lectures. Sierpiński's lecture was devoted to the problem of the role and meaning of the concept of correspondence in mathematics (Sierpiński 1909) and Janiszewski's lecture to the controversy between realism and idealism in the philosophy of mathematics (Janiszewski 1916).

The interests of Polish mathematicians for logic and the philosophy of mathematics as well as their conviction concerning the importance of those disciplines for mathematics manifested itself in many ways, as for instance in the publication in 1915 of a *Handbook for Autodidacts* (*Poradnik dla samouków*) devoted to mathematics. Among the authors who contributed were Jan Łukasiewicz who wrote the introductory chapter "On Science", W. Sierpiński who wrote the one on set theory and Z. Janiszewski who was the main contributor and the soul of the

R. Murawski (✉)
Adam Mickiewicz University, Poznań, Poland
e-mail: rmur@amu.edu.pl

S. Lapointe et al. (eds.), *The Golden Age of Polish Philosophy*, Logic, Epistemology, and the Unity of Science 16, DOI 10.1007/978-90-481-2401-5_9,

whole enterprise. Janiszewski wrote the chapters on the foundations of geometry, on mathematical logic and on philosophical issues in mathematics. The latter is a wide-ranging presentation of the main problems and views in the philosophy of mathematics, with an extensive bibliography in which the then current positions and literature were well represented. The third volume of the manual (published in 1923) comprised a chapter "On the meaning of mathematical logic for mathematics" by Jan Sleszyński. In his essay, Sleszyński expressed the belief that mathematical logic, because it is an autonomous discipline, cannot be reduced to the methodology of mathematics and he defended logic against various objections formulated by mathematicians and philosophers (among others, by Poincaré).[1]

Connections between mathematicians and logicians in Warsaw were very good and they collaborated closely. Both groups saw mathematical logic and the methodology of mathematics as disciplines that are autonomous from both mathematics and philosophy. And both groups were convinced that those disciplines play a fundamental role in the development of mathematics. They believed that mathematics and mathematical logic should be neutral as regards various philosophical controversies, and that they should be developed independently of any philosophical presuppositions. For example, many Polish mathematicians were convinced that the philosophical concerns relative to the axiom of choice must be kept separated from the questions concerning the role of the latter in mathematics. Sierpiński wrote:

> Still, apart from our personal inclination to accept the axiom of choice, we must take into consideration, in any case, its role in set theory and in calculus. On the other hand, since the axiom of choice has been questioned by some mathematicians, it is important to know which theorems are proved with its aid and to realize the exact point at which the proof has been based on the axiom of choice; for it has frequently happened that various authors have made use of the axiom of choice in their proofs without being aware of it. And after all, even if no one questioned the axiom of choice, it would not be without interest to investigate which proofs are based on it and which theorems are proved without its aid — this, as we know, is also done with respect to other axioms. (1965, 95)

What Sierpiński means is that when investigating (controversial) axioms one ought, using any fruitful method, to treat them as purely mathematical constructions abstract from philosophical controversies (which are to be considered as "private" matters).

---

[1] In Cracow, the other "centre" for mathematics in Poland in the interwar period, the attitude towards logic and philosophy was somewhat different. The latter finds an illustration in the well-known controversy between Stanisław Zaremba and Jan Łukasiewicz on the concept of magnitude (which in fact concerned the role of logic in mathematics) — cf. Murawski (2004). Zaremba represented the view that logic is a part of mathematics whereas Łukasiewicz considered (mathematical) logic as an autonomous discipline providing the foundations and methodology of mathematics. The leaders and founders of the Polish Mathematical School in Warsaw accepted Łukasiewicz' position. They stressed the role of set theory and of mathematical logic and considered logic to be at the center of mathematics (Zaremba saw its place in the periphery of mathematics).

Paradoxically, one of the consequences of the fraternal attitude between Polish logicians and mathematicians is that none of them ever attempted to develop a comprehensive philosophy of mathematics and logic (Stanisław Leśniewski and Leon Chwistek are here the exceptions!). They formulated their philosophical opinions concerning mathematics or logic only occasionally, and only with respect to problems they happened to find interesting or on which they happened to be working. Hence, there were in Poland no genuine philosophers of mathematics: philosophical remarks were formulated by logicians and mathematicians only at the margin of their proper mathematical or logical works (and were not meant to be results in themselves).

The then current trends and views in the philosophy of mathematics, i.e., logicism, intuitionism and formalism, were of course well known (a number papers discussing those tendencies, their meaning and development were published), but none of them was represented in the Warsaw School. Moreover, the Warsaw School did not represent any other trend: it had no official philosophy of logic and mathematics. This can be seen as the upshot of the belief that philosophy is autonomous from logic and mathematics. Opinions in the field of the philosophy of logic and mathematics were treated as "private" problems and philosophical declarations were reluctantly and seldom made – and whenever they were, one invariably stressed implicitly or explicitly that these were personal opinions.

Although some logical investigations were motivated by philosophical problems (e.g. Łukasiewicz's many-valued logics) formal logical constructions were always kept separate from their philosophical interpretations. This attitude was strengthened by Alfred Tarski (1901–1983) and Andrzej Mostowski (1913–1975) who both claimed that a logician or a mathematician can have philosophical views or sympathies quite different from those which could be suggested by the range of problems on which she is working. They provided good examples of this approach in their own work. Mostowski wrote about Tarski:

> Tarski, in oral discussions, has often indicated his sympathies with nominalism. While he never accepted the "reism" of Tadeusz Kotarbiński, he was certainly attracted to it in the early phase of his work. However, the set-theoretical methods that form the basis of his logical and mathematical studies compel him constantly to use the abstract and general notions that a nominalist seeks to avoid. In the absence of more extensive publications by Tarski on philosophical sub ects, this conflict appears to have remained unresolved.
>
> (Mostowski, 1967, 81)

Mostowski himself was a partisan of constructivism but his logical and foundational investigations did not take into account the corresponding methodological restrictions. There are other examples. Some, among others by Tarski, carried out investigations in intuitionistic logic without accepting intuitionism and we find in the programme of Janiszewski and the Polish Mathematical School set-theoretical foundations for mathematics in the methodological, but not philosophical sense.[2]

---

[2] Note that in Janiszewski's manifesto — cf. Janiszewski (1918) — one finds nowhere the words "set theory". In fact Janiszewski deliberately avoided it — it was a sign of caution. The volume of *Nauka Polska* in which Janiszewski's paper appeared was published just at the culminating point

8.2 Philosophies of Mathematics

What were the philosophical positions of Polish logicians, philosophers and mathematicians? Let us start with their position towards psychologism. According to this positions which was popular in the philosophy of logic and mathematics in the late 19th century, the objects studied by logic and mathematics are mental entities and come to be known just like other mental facts. Twardowski was the first Polish philosopher to take a step towards antipsychologism. The next step was taken in a paper by Łukasiewicz "Logika a psychologia" (Logic and psychology) (1907) in which he declared himself firmly against psychologism in logic. His objections consisted in claiming that (1) while logical laws are certain, psychological laws (being in fact empirical) are only probable, that (2) the laws of logic and the laws of psychology differ in content: the former concern the connections between the truth and falsehood, and the latter describe relationships between psychic phenomena, that (3) the terms "thinking" and "judgement" have different meaning in psychology and in logic. Łukasiewicz concludes (1907, 65):

> The clarification of the relationship between logic and psychology may prove to be to the advantage of both disciplines. Logic will be purified of the weeds of psychologism and empiricism, which hamper its true development, and the psychology of cognition will rid itself of the elements of apriorism behind which the genuine light of its truth could not fully show itself. It must be borne in mind that logic is an *a priori* science, like mathematics, while psychology, like any natural science, is, and must be, based on experience.

Łukasiewicz's arguments against psychologism were similar to those of Husserl and Meinong. They were universally accepted in Poland. As a consequence, virtually everyone became convinced that the certainty of the theorems of logic cannot be explained by psychological arguments. Antipsychologism was however a paradoxical solution to the problem of the certainty of logic: since almost all Polish logicians were sympathizers of genetic empiricism, an aprioristic solution to this problem could hardly have been accepted.

Polish logicians did not accept the concept of logic as a pure syntax, a view, which was popular at the time and which had been developed under the influence of Hilbert's metamathematics and the philosophy of language of the Vienna Circle. Chwistek who treated his semantic systems as formal systems of expressions is here the exception (On Chwistek, see Linsky, this volume). The Warsaw School adopted a semantic approach. It is with this as a background that one should consider Tarski's attempt at a semantic foundations of mathematics in the 1930s. The same holds for Leśniewski's original approach to the problem of the nature of logic which is usually referred to as "intuitive formalism": while engaged in the task of carrying out the formalization of logical systems, Leśniewski claimed that formal expressions always encode fixed intuitive contents. At any rate, the semantic approach implied the rejection of the "analytical" conception of logic, i.e., the thesis that logic is a

of the dispute between Zaremba and Łukasiewicz and in the same volume the paper by Zaremba was published as well.

collection of tautologies whose content is empty. Leśniewski (as did Kotarbiński) claimed, for instance, that logic describes the most general features of being and, thus, that it plays the role of a general theory of the real world. Tarski, on his part, described the concept of tautology as vague and, as we will see in more detail below, did not see any objective basis for the division of terms into logical and extra-logical. Andrzej Mostowski, epitomizing the spirit of this approach, wrote that various metamathematical results

> obtained by the mathematical method confirm therefore the assertion of materialistic philosophy that mathematics is in the last resort a natural science, that its notions and methods are rooted in experience and that attempts at establishing the foundations of mathematics without taking into account its originating in natural sciences are bound to fail. (1955, 42)

And added:

> An explanation of the nature of mathematics does not belong to mathematics but to philosophy, and is possible only within the limits of a broadly conceived philosophical view treating mathematics not as detached from other sciences but taking into account its being rooted in natural sciences, its applications, its associations with other sciences and, finally, its history. (1955, 42)

Łukasiewicz's views concerning the problem of the relation between logic and mathematics, on the one hand, and reality on the other were not unswerving. In "Creative Elements in Science" he claimed that logical and mathematical judgments are *a priori* truths about a world of ideal entities. (1912, 73–74) Hence, he treated both disciplines as unrelated to experience. The discovery of many-valued logics led him however to maintain that logical systems can be given an ontological interpretation and that experience can help to decide which systems of logic is fulfilled in reality (1936, 199). Later on, he leaned towards conventionalism and relativism:

> We have no means to decide which of the n-valued systems of logic [...] is true. Logic is not a science of the laws of thought or of any real object; it is, in my opinion, only an instrument which enables us to draw asserted conclusions from asserted premises. [...] The more useful and richer a logical system is, the more valuable it is. (1952, 208)

Kazimierz Ajdukiewicz (1890–1963) was also an advocate of conventionalism (in a radical version). He claimed (e.g., in 1934) that logic is something implied by meaning rules (the rules of sense), both axiomatic and deductive ones. He later on (e.g., in 1947) abandoned radical conventionalism and claimed that the laws of logic refer indirectly to experience and that they should be treated as rules of inference: they belong to metascience and are mainly of a methodological character.

Nominalism was vastly discussed and commented upon. Leśniewski was a declared nominalist and consequently denied the existence of general objects. He claimed that the logical systems he created consisted of a finite number of individual inscriptions. More on this in what follows. Kotarbiński in the early and radical version of his reism held views similar to those of Leśniewski. Tarski had nominalistic leanings (inherited from Leśniewski) as well, but the needs of metamathematics led him to abandon these sympathies. In particular, while Tarski treats formulas themselves as mere physical entities, his semantics appeals to formula types, that is, classes of equiform formulas.

Łukasiewicz's attitude towards nominalism was different, and mostly negative. He maintained that Tarski's arguments in defense of nominalism were not sufficient. He thought that logicians merely *used* nominalistic terminology, but were not in fact fully-fledged nominalists. He was inclined to interpret logic in an outright neo-Platonist spirit:

> In concluding these remarks I should like to outline an image which is connected with the most profound intuitions which I always experience in the face of logistic. That image will perhaps shed more light on the true background of that discipline, at least in my case, than all discursive description could. Now, whenever I work on even the least significant logistic problem — for instance, when I search for the shortest axiom of the implicational propositional calculus — I always have the impression that I am facing a powerful, most coherent and most resistant structure. I sense that structure as if it were a concrete, tangible object, made of the hardest metal, a hundred times stronger than steel and concrete. I cannot change anything in it; I do not create anything of my own will, but by strenuous work I discover in it ever new details and arrive at unshakable and eternal truths. Where is and what is that ideal structure? A believer would say that it is in God and His thought. (1937, 13)

Note that Łukasiewicz – Ajdukiewicz and Czeżowski held similar opinions – stressed that logic itself cannot solve the philosophical controversy over universals. Nonetheless, claims to the effect that logic is nominalistic are, according to Łukasiewicz, groundless.

Leśniewski held the chair of philosophy of mathematics at the University of Warsaw (from 1919 till his death in 1939). Although he was convinced that philosophical investigations are hopeless and never lead to definite solutions, he himself was not deprived of a philosophical approach to logic. His aim was to construct a system of logic – Leśniewski maintained that logic should be extensional and bivalent – that would satisfy two general requirements: it should serve as the foundation for mathematics, and it should be constructed so as to be free of all ambiguities. His investigations led him to construct three systems: Protothetic, Ontology and Mereology. (On Leśniewski's systems, see Betti, Miéville, this volume) Leśniewski conceived of these logical systems in a nominalistic way. Language for him is a collection of concrete individual inscriptions. Only expressions which have actually been written exist – and he admitted no "potential" existence of any sort. Leśniewski called this position "constructive nominalism", and the latter is connected with the view that formal expressions encode intuitive contents, that is, "intuitive formalism". Indeed, one consequence of his position is the rejection of the idea that logic and mathematics are games using symbols that are devoid of meaning. According to Leśniewski every language system says "something about something", is a way of expressing what is intuitively true and is an indispensable way of encoding and transmitting logical intuitions.

Chwistek is known mainly for his logical works, i.e., for his simplification of Russell and Whitehead's theory of types of (which he carried out in a nominalistic spirit). The aim of his logical investigations was to create a comprehensive system of logic and mathematics based on a theory of expressions (which he called "rational metamathematics"). His results and ideas had rather limited influence. The main reason for this is the complicated and nonstandard notation he – too – used when presenting his results. He conceived of his position as a form of "critical" rationalism. According to him there are two sources of knowledge: experience and deduction. The aim of science is to use mathematical expressions to describe the objects given in experience. More generally, mathematical formulas are merely descriptions of experiences and cannot be treated as laws concerning objects which are not given by experience. Presumably for this reason, he believed that the methods used in science and in philosophy should be constructive. Indeed, one of Chwistek's best known philosophical realization is his theory of multiple realities. It was published for the first time in his paper from 1917 "Trzy odczyty odnoszą ce sieę do pojeę cia istnienia" (Three Lectures concerning the Concept of Existence) and found its penultimate presentation in *Granice nauki* (The Limits of Science, 1935). In this book, Chwistek postulates four types of reality – the reality of impressions, the reality of images, the reality of things and physical reality (constructed by science) – and attempts to characterize the properties of those types of realities by a suitable sets of axioms.

Chwistek was both for nominalism and against formalism. He claimed that the objects of deductive systems are expressions. But, on his view, geometry is an experimental discipline. He considered the development of non-Euclidean geometries to be one of the most important achievements in science: it refuted Kantian idealism and the view that geometry is given *a priori*. Geometry, mathematics as a whole, as well as the other sciences, should be developed constructively, i.e., one should base them on axioms and definitions such that the theorems deduced from them are in accordance with experience. This, one might want to argue, suggests that he would have accepted conventionalism. But in *Granice nauki* he clearly rejected it. Not only did he claim that conventionalism is incorrect as far as the natural sciences are concerned, he also maintained that it is a source of abuse in the social sciences (since it reduces truth to usefulness and efficiency and, in so doing, leads to the reinforcement of the ruling class).

8.3 Tarski

In the two previous sections we mentioned some of Tarski's views in the philosophy of logic and mathematics. Tarski's role in the development of mathematical logic and the foundations of mathematics was however enormous, so let us say more about his philosophy (cf. also Woleński 1989, 1993 and 1995, and this volume, Murawski/Woleński forthcoming).

Tarski was interested in philosophical problems and very actively participated in the philosophical life of his time. He was convinced of the philosophical significance of his works, in particular of his work on truth. In (1933) he wrote:

> I shall be satisfied if this paper convinces the reader that the method used above is a device which may be indispensable when considering purely philosophical problems. [...] The central problem of this paper — constructing of a definition of truth for sentences and providing a scientific basis for the theory of truth — belongs to epistemology and is one

> of the main problems in this domain of philosophy. Hence I expect that specialists in epistemology will take an interest in it, that — not allowing themselves to be discouraged by difficult notions and methods that have not so far been applied in this field — they will critically examine the results contained in it and will be able to use them in their further studies. (1933, 115)

He described himself as:

> [...] a mathematician (as well as a logician, perhaps a philosopher of a sort) [...]
> (Tarski, 1944, 374)

Tarski's philosophical attitude was typically antimetaphysical. He supported the idea of a scientific philosophy and, in particular, adopted a programme which aimed at detailed and systematic analyses of philosophical concepts. Such a philosophy was meant to be minimalistic, anti-speculative and skeptical towards a number of fundamental traditional philosophical problems. Tarski most likely inherited this attitude from the Lvov-Warsaw School and the latter was strengthened through his contacts with the Vienna Circle. He also advocated empiricism and – for reasons however utterly different from those of his successors such as Quine – rejected the analytic/synthetic distinction, claimed that the concept of a tautology is unclear and stressed that logical and empirical truths belong to the same generic category. One could therefore argue that he did not consider the boundaries between formal and empirical disciplines to be perfectly defined.

Influenced by Leśniewski and Kotarbiński he was inclined to a rather strongly nominalistic conception of linguistic expressions. He treated sentences as concrete physical objects and languages as consisting of token-expressions, though metalogical studies forced an understanding of them as expression-types. Tarski sharply contrasted colloquial, natural language and formalized language.

Tarski was also inclined to identify mathematics with the deductive method. He maintained that there is no hard demarcation between formal and empirical sciences and he admitted that logical and mathematical theories could be rejected on empirical grounds. Those were his "private" philosophical views and Tarski was at least consistent – whether he was correct is another issue – in claiming that the latter did not influence his logical and mathematical researches and that, as a result, his logical and mathematical theories were independent of any philosophical presuppositions. In the paper "Über einige fundamentale Begriffe der Methodologie der deduktiven Wissenschaften" he explicitly claimed that:

> [...] no particular philosophical standpoint regarding the foundations of mathematics is presupposed in the present work. (1930, 362)

This was indeed typical of the Warsaw School's approach to logic as a whole. The idea that logical and mathematical studies are independent from philosophical views explains the cognitive conflict and discrepancy between Tarski's nominalistic and empiricist sympathies and his "Platonism" in mathematical and logical practice. This attitude at any rate enabled him to contribute to various important foundational streams without having to accept their philosophical assumptions or attempting to reconcile his research with his underlying philosophical commitments. Tarski summarized his programme in metamathematics in the following words:

> As an essential contribution of the Polish school to the development of metamathematics one can regard the fact that from the very beginning it admitted into metamathematical research all fruitful methods, whether finitary or not. (1954, 18)

Note that Tarski's attitude also fully agrees with that of Polish mathematicians, as we've indicated above. According to the latter, one may examine problems using any fruitful methods and should make no philosophical assumption. There is no need to make explicit one's philosophical views concerning the problems investigated; this does not belong to one's scientific duties. It is in this sense that philosophy a "private" affair.

8.4 Conclusion

Both Polish logicians and mathematicians believed that philosophical problems in logic and mathematics are important. They knew the current views and trends in the philosophy of mathematics quite well, commented upon them and proposed a number of views of their own. But on the other hand they treated logic and mathematics as disciplines independent of the philosophical reflection on them, and indeed independent of any philosophical presuppositions. They sharply separated mathematical and logical research practice and philosophical discussions concerning logic and mathematics. Philosophical views and opinions were treated as "private" matter that should not influence the mathematical and metamathematical investigations, where all correct methods can and should be used.

Acknowledgments We wish to acknowledge the financial support of the Foundation for Polish Science

References

Ajdukiewicz Kazimierz (1934), 'Das Weltbild und die Begriffsapparatur', *Erkenntnis* 4, 259–287.

Ajdukiewicz, Kazimierz (1978), 'Logika a doświadczenie' (Logic and Experience), *Przegląd Filozoficzny* 43 (1947), 3–22. Eng. trans. 'Logic and Experience', in K. Ajdukiewicz. *The Scientific World-Perspective and Other Essays (1931–1963)*, J. Giedymin (ed.), D. Reidel, Dordrecht.

Chwistek, Leon (1917), 'Trzy odczyty odnoszą ce się do pojęcia istnienia' (Three Lectures concerning the Concept of Existence), *Przegląd Filozoficzny* 20, 122–151.

Chwistek, Leon (1935), *Granice nauki. Zarys logiki i metodologii nauk ścisłych (Limits of Science* An Outline of Logic and Methodology of Science), Książnica-Atlas, Lwów-Warszawa. Eng. trans. *The Limits of Science. An Outline of Logic and Methodology of Science*, New York–London, The International Library of Philosophy, 1948.

Janiszewski, Zygmunt (1918), 'O potrzebach matematyki w Polsce' (On the needs of mathematics in Poland), *Nauka Polska*, Warszawa, 11–18.

Janiszewski, Zygmunt (1915), 'Zagadnienia filozoficzne matematyki' (Philosophical problems of mathematics), in *Poradnik dla samouków. Wskazówki metodyczne dla studiujących*

poszczególne nauki, vol. 1, Wydawnictwo A. Heflicha i St. Michalskiego, Warsaw, 462–489.
Janiszewski, Zygmunt (1916), 'O realizmie i idealizmie w matematyce' (On realism and idealism in mathematics), *Przegląd Filozoficzny* 19, 161–170.
Łukasiewicz, Jan (1907), 'Logika a psychologia' (Logic and psychology), *Przegląd Filozoficzny* 10, 489–491.
Łukasiewicz, Jan (1912), 'O twórczości w nauce' (Creative elements in science), in *Ksieęga pamiątkowaku uczczeniu 250 rocznicy założąenia Uniwersytetu Lwowskiego*, Lwów, 1–15. Eng. trans. 'Creative elements in science', in *J. Łukasiewicz. Selected Works*, L. Borkowski (ed.), North-Holland/PWN, Amsterdam/ Warsaw 1970, 1–15.
Łukasiewicz, Jan (1936), 'Logistyka a filozofia' (Logistic and philosophy), *Przegląd Filozoficzny* 39, 115–131. Eng. trans. 'Logistic and philosophy', in *J. Łukasiewicz. Selected Works*, L. Borkowski (ed.), North-Holland/PWN, Amsterdam/Warsaw 1970, 218–235.
Łukasiewicz, Jan (1937) 'W obronie logistyki' (In defence of logistic), *La pensée catholique et la logique moderne*, Wydawnictwo Wydziału Teologicznego UJ, Kraków, 7–13. Eng. trans. 'In defence of logistic', in J. Łukasiewicz. *Selected Works*, L. Borkowski (ed.), North-Holland/PWN, Amsterdam/Warsaw 1970, 236–249.
Łukasiewicz, Jan (1952), 'On the intuitionistic theory of deduction', Indagationes Mathematicae 14, 202–212. Reprinted in: in J. Łukasiewicz. *Selected Works*, L. Borkowski (ed.), North-Holland/PWN, Amsterdam/Warsaw 1970, 325–340.
Mostowski, Andrzej (1967), 'Tarski, Alfred', in P. Edwards (ed.) *The Encyclopedia of Philosophy*, vol. 8, Macmillan, New York, 77–81.
Mostowski, Andrzej (1955) (in collaboration with A. Grzegorczyk, S. Jaśkowski, J. Łoś, S. Mazur, H. Rasiowa and R. Sikorski) 'The present state of investigations on the foundations of mathematics', *Rozprawy Matematyczne* 9, 1–48.
Murawski, Roman (2004), 'Philosophical reflection on mathematics in Poland in the interwar period', *Annals of Pure and Applied Logic* 127, 325–337.
Murawski, Roman, Woleński, Jan (forthcoming),'Tarski and his Polish predecessors (on truth)', in Douglas Patterson (ed.), *Alfred Tarski: Philosophical Background, Development and Influence*, Oxford, Oxford University Press.
Sierpiński, Wacław (1909), 'Pojęę cie odpowiedniości w matematyce' (The concept of correspondence in mathematics), *Przegląd Filozoficzny* 12, 8–19.
Sierpiński, Wacław (1965), *Cardinal and Ordinal Numbers*, PWN, Warsaw.
Sleszyński, Jan (1923), 'O znaczeniu logiki dla matematyki' (On the meaning of logic for mathematics), in *Poradnik dla samouków, vol. 3: Matematyka. Uzupełnienia do tomu pierwszego*, Wydawnictwo A. Heflicha i St. Michalskiego, Warsaw, 39–52.
Tarski, Alfred (1930), Fundamentale Begriffe der Methodologie der deduktiven Wissenschaften. I', *Monatshefte für Mathematik und Physik* 37, 361–404.
Tarski, Alfred (1933), *Pojęcie prawdy w językach nauk dedukcyjnych* (The notion of truth in languages of deductive sciences), Nakładem Towarzystwa Naukowego Warszawskiego, Warsaw. Eng. Trans. 'The concept of truth in formalized languages', in: *Logic, Semantics, Metamathematics: Papers From 1923 to 1938 by Alfred Tarski*, Oxford, Oxford University Press, 1956; second edition with corrections and emendations, J. Corcoran (ed.), Indianapolis, Hackett, 1983, 152–278.
Tarski, Alfred (1944) 'The semantic conception of truth and the foundations of semantics', *Philosophy and Phenomenological Research* 4, 341–375.
Tarski, Alfred (1954), 'Contribution to the discussion of P. Bernays "Zur Beurteilung der Situation in der beweistheoretischen Forschung"', *Revue Internationale de Philosophie* 8, 16–20.
Woleński, Jan (1989), *Logic and Philosophy in the Lvov-Warsaw School*, Dordrecht, Kluwer.
Woleński, Jan (1993) 'Tarski as a Philosopher', in F. Coniglione, R. Poli and J. Woleński (eds.),*Polish Scientific Philosophy: The Lvov-Warsaw School*, Editions Rodopi, Amsterdam–Atlanta, GA, , 319–338.
Woleński, Jan (1995), 'On Tarski's Background', in ed. J. Hintikka (ed.), *Essays on the Development of the Foundations of Mathematics*, Dordrecht, Kluwer, 331–341.

Chapter 9
Tarski's Engagement with Philosophy

Paolo Mancosu

9.1 Introduction

Tarski's legacy to contemporary philosophy is huge. His epoch making works on truth and logical consequence are still widely discussed and represent key reference points in the area of history and philosophy of logic, philosophy of language and philosophy of science. Yet, however charged with philosophical potential, his papers cannot be said to provide a picture of Tarski as a general philosopher. When it comes to philosophical views, Tarski was extremely reticent to put anything in print and even those who knew him well, such as Mostowski, were unable to distil much out of his published writings. For instance, in the entry "Tarski" for the *Encyclopedia of Philosophy* Mostowski, summarizing Tarski's engagement with philosophy, writes:

> Tarski, in oral discussions indicated his sympathies with nominalism. While he never accepted the "reism" of Tadeusz Kotarbiński, he was certainly attracted to it in the early phase of his work. However, the set-theoretical methods that form the basis of his logical and mathematical studies compel him constantly to use the abstract and general notions that a nominalist seeks to avoid. In the absence of more extensive publications by Tarski on philosophical subjects, this conflict appears to have remained unresolved"
>
> (Mostowski, 1967, 81)

Tarski himself joked about this on the occasion of a series of comments he made in an ASL meeting in Chicago in 1965. There, talking about his anti-Platonism, he said:

> I happen to be, you know, a much more extreme anti-Platonist. [...] However, I represent this very [c]rude, naïve kind of anti-Platonism, one thing which I could describe as materialism, or nominalism with some materialistic taint, and it is very difficult for a man to live his whole life with this philosophical attitude, especially if he is a mathematician, especially if for some reasons he has a hobby which is called set theory (p. 3, Transcript of remarks, ASL meeting, Chicago, Illinois, April 29, 1965, Bancroft Library; now in Rodriguez-Consuegra 2007)

P. Mancosu (✉)
University of California Berkeley, Berkeley, CA, USA
e-mail: mancosu@socrates.berkeley.edu

S. Lapointe et al. (eds.), *The Golden Age of Polish Philosophy*, Logic, Epistemology, and the Unity of Science 16, DOI 10.1007/978-90-481-2401-5_10,

Jan Woleński has devoted two papers (Woleński 1993, 1995) to Tarski as a philosopher and they are up to now the best treatment available of Tarski's philosophical views (but see also Suppes 1988 and Rojszczak 2002). In Woleński 1993 he correctly distinguishes "between one's philosophy and the philosophical consequences or implication of one's works" and then concentrates on Tarski's explicit philosophical views and reports on (1) Tarski's philosophical biography; (2) Tarski's views in metaphilosophy; (3) Tarski's philosophy of language; (4) Tarski's views on truth; (5) Tarski's philosophy of science, including some problems in the philosophy of logic and mathematics; (6) some of Tarski's general views on science and society.

Woleński (321–22) remarks that "Tarski neither created a philosophical system nor even presented a part of philosophy in a systematic way". However, he is right in pointing out that Tarski's philosophical views are important in that "omitting Tarski's philosophy and his philosophical activity would mean an impoverishment of his scientific profile and the role which he played in the development of contemporary philosophy".

I will take the value of Woleński's work for granted and try to go further. One of the first things to remark about Woleński's treatment is that it is based almost exclusively on quotations from published sources (and, as I said, he does a great job with it). Tarski's philosophical engagement seems to have been stronger before his move to Berkeley in 1942 and his archive at Berkeley is rather disappointing from this point of view. This is not surprising given the circumstances surrounding Tarski's move to the USA. In 1939 he was in Harvard for a conference and a series of talks when the Germans invaded Poland. He stayed in the United States and his family joined him only in 1945 (see Feferman and Ferman 2004 on the details). We have to conjecture that most of the correspondence, papers etc. that he had in his household in Warsaw were destroyed or lost in the vicissitudes of the war. The scientific loss was great. Let me give just two examples. In a letter to Neurath dated 28.4.36 Tarski claims to be very busy since a very large mathematical book that he is co-authoring with Lindenbaum should soon appear. As it transpires from a letter to Popper (dated 1.2.36) this was a book on set theory (which certainly went back to a joint article Tarski and Lindenbaum wrote in 1926). Moreover, we know from a postcard from Hempel to Neurath (dated 18.9.38) that Tarski in 1938 had written an article on the significance of the concept of truth:

> So Tarski visited you today; there must have been much to discuss; I recommended that he give you his manuscript on the significance of the concept of truth; hopefully he did so. It seems to me to be very good that the question be explained in a more general form and also independently of the computational aspects; it would be something good for *Erkenntnis*.[1]

---

[1] Heute ist nun Tarski bei Ihnen gewesen; sicher gabs viel zu diskutieren; ich schlug ihm vor, Ihnen sein MS über die Bedeutung des Wahrheitsbegriffs zu geben; hoffentlich hat ers getan. Es scheint mir sehr gut zu sein, dass die Frage auch in allgemeinerer Form und auch ausserhalb des Kalkülbare-Problemgebiets erörtert wird; das wäre was Schönes für die Erkenntnis. (Hempel to Neurath, 18.9.38)

Tarski discussed with Neurath the possibility of publishing this manuscript in *Erkenntnis* during his visit to Holland in September 1938 but nothing came of it. How close would it have been to his 1944 article on truth? We'll probably never know.

In addition to his unpublished writings we also regret the loss of a great part of the scientific correspondence before the Berkeley period, correspondence that is remarkably absent from the Berkeley archive. This leads me to the main part of my paper.

Before his move to the USA, Tarski was heavily involved with two major scientific groups. First of all with the Polish logicians and philosophers, who are known under the name of Lvov-Warsaw school. In this case, he had ample opportunity to discuss things in person and the (partial) lack of scientific correspondence might not be due to a loss on account of the war but rather only reflects the fact that there was no need to correspond because of the physical proximity of the members of the Lvov-Warsaw school. In any case, because of the language barriers, it would be up to Polish speaking scholars to find out whether in the archives of several members of the Lvov-Warsaw school we might find Tarskian materials and, of course, to evaluate their importance.

The second group which played a very important role in Tarski's scientific work was the Vienna Circle and the Unity of Science movement.[2] I suggest that in this case there is ample evidence that our knowledge of Tarski's scientific biography and philosophical views can be deepened by a systematic search in the archives of the various members of the Vienna circle. (By the way, I should add that the situation here is also true for other members of the Lvov-Warsaw school.)

I would like to show the potential pay-off of such investigations by reporting on three episodes (dating from the mid-1930s to early 1940s) from Tarski's scientific/philosophical life about which the archives of members of the Vienna circle provide interesting and valuable information. The first of them is the correspondence between Tarski and Neurath; the second is conveyed in a report on Tarski's views written by Hempel in a letter to Neurath; finally, the third, and perhaps most exciting, comes from eighty pages of notes from the Carnap *Nachlass* concerning the meetings at Harvard in 1940–1941 between Tarski, Carnap and Quine on the project of a nominalistic foundation of mathematics and science. To keep this article within reasonable length, I have decided to proceed impressionistically, that is, I will provide only samples of the materials I will draw your attention to. In a series of papers (Mancosu 2005, 2006, 2008) I treat these issues with the care and length they deserve.

---

[2] This should be taken broadly to include also thinkers such as Popper and Woodger who were not part of the Vienna Circle but whose interests overlapped with the goals of the Unity of Science movement. There is extensive correspondence between Tarski and Popper and between Tarski and Woodger (see list of archival resources at the end of the paper). A first step in the direction of studying the relationship between Tarski and the Vienna Circle is Woleński, Köhler, 1998. However, almost none of the material I am appealing to has been discussed there.

9.2 Neurath and Tarski on the Metaphysical Import of Semantics

The lasting impact of Tarski's theory of truth on Carnap's thought is well known.[3] What is usually paid less attention to is that Tarski's work appeared in the midst of a debate on the nature of truth which had divided the Vienna circle.[4] Central contributions to this debate were Schlick's article on 'On the foundation of knowledge'(1934), Neurath's reply 'Radical Physicalism and the Real World' (1934), and Hempel's 'On the Logical Positivists' Theory of Truth'(1935). The immediate impact of Tarski's investigation on truth is evident in successive contributions related to this debate such as Carnap's 'Truth and confirmation' (1936) and Lutman-Kokoszyńska's 'On the Absolute Concept of Truth and some other Semantical Concepts' (1936). The most important event in this connection is the Paris congress of 1935 in which Tarski delivered a talk presenting his work on semantics and the theory of truth (in addition to a different talk on logical consequence). According to Ayer, "the highlight of the Congress was the presentation by Tarski of a paper summarizing his theory of truth" (Ayer, 1977, 116). The reason why Tarski's talk produced such lively interest is that it seemed to provide a rigorous account of the position according to which truth is defined as correspondence between language and reality. A glimpse of the discussion following Tarski's presentation can be obtained by reading Neurath's long overview of the Congress published in *Erkenntnis* (Neurath 1936) At the Paris congress, Neurath had proposed an alternative view of truth:

> From the point of view of terminology he [Neurath] thinks that one should reserve the use of the term "true" for that Encyclopedia, among the many consistent ones which are controlled by protocol sentences, that has been chosen, so that each consequence of this Encyclopedia and each new sentence accepted into it would be called "true" and any one contradicting it would be called "false".
>
> (Neurath 1936, 400)[5]

This was simply a restatement of a view that Neurath had previously defended and that had led Schlick to criticize, in his 1934 article on the foundations of knowledge, Neurath's account of truth as a "coherence" theory of truth. While Neurath's

---

[3] See Carnap's *Autobiography* and *Introduction to Semantics* (1942, x). Moreover, Tarski was instrumental in making improvements to the English edition of *Logische Syntax der Sprache*. We know from a letter from Kokoszyńska to Carnap (dated 24.VII.35) that Tarski had sent Carnap a list of corrections along with a letter. Unfortunately, neither the list of corrections nor the letter seem to be in the Carnap *Nachlass*. The letter from Kokoszyńska, where incidentally she also says she is sending her own list of corrections, is classified under RC 088-57-14.

[4] See Mancosu 2008, Hofmann-Grüneberg 1988 and Uebel 1992. For an earlier account see Tugendhat 1960.

[5] Vom terminologischen Standpunkt aus meine er, daß man den Terminus "wahr" für diejenige unter den vielen in sich widerspruchslosen, durch Protokollsätze kontrollierten Enzyklopädien reservieren könnte, für die man sich entschieden hat, so daß "wahr" jede Konsequenz dieser Enzyklopädie und jeder neue in sie aufgenommener Satz gennant würde, "falsch" jeder mit ihr in Widerspruch tretende Satz. (Neurath, 1936, 400)

opposition to Tarski's theory of truth (and that presented by Lutman-Kokoszyńska, which was derivative on Tarski's definition of truth) can be detected from Neurath's report, a full picture of the range of Neurath's motives and of the context in which Neurath's objections emerged can only be obtained through a detailed study of the epistolary exchange that Neurath had on this subject with Tarski, Carnap, Lutman-Kokoszyńska and Hempel.[6]

Since our focus is on Tarski, let me restrict attention to the correspondence between Neurath and Tarski. It consists of 41 letters from Tarski and 42 from Neurath spanning the period 1930–1939. In 1992, Haller published three of these letters (Haller 1992). The letters published by Haller are of great interest for clarifying certain historical matters related to the chronology and mutual influences between the Polish philosophers and the Vienna Circle. However, I will focus on another aspect of the correspondence. The letters of interest for us date from the period after the Paris Congress of 1935. I will only provide some passages concerning the definition of truth, one giving voice to a compact summary of Neurath's doubts and then a lengthy reply by Tarski. On April 24, 1936, Neurath writes:

> I thank you very much for your reflections on our "truth definitions". Of course there are to begin with only terminological differences but I have the strong impression that in the discussion concerning the domain of the real sciences your intuition slips very easily into metaphysics. One should fully speak one's mind on this issue. I wrote to Dr Lutman-Kokoszyńska about this. When you hold that it is trivial to say that one speaks with the language about the language then I can only rejoin that an essential part of science consists in defending trivialities against errors. From the beginning of the Vienna Circle, for instance, I have fought against Wittgenstein's attempt to introduce a sort of "elucidations" and thus "illegitimate", almost non- or pre-linguistic considerations in order to then speak of the opposition between "the" language and "the" reality, and hence to speak outside the language.[...] And insofar as your terminological choice suggests objectionable consequences, it has perhaps not come about independently of these consequences. On the one hand, one emphasizes that this concept of truth holds only for formalized languages. On the other hand, the concept of truth is of practical interest precisely in non formalized domains. For this reason, if one is not simply to get rid of the term, I am in favor of my terminology, for the latter remains applicable also in non formalized domains. By contrast the terminology you and Lutman use leads to bad things when it is applied to non formalized domains.
>
> (Neurath to Tarski, 24.IV.36)[7]

---

[6] I exploit these materials in Mancosu 2008. The correspondence between Neurath and Kokoszyńska contains 19 letters from Kokoszyńska and 14 letters from Neurath. They span the period 1934–1939.

[7] Ich danke Ihnen für die Mitteilungen über unsere "Wahrheitsdefinitionen". Natürlich liegen zunächst nur terminologische Unterschiede vor, aber ich habe sehr den Eindruck, dass bei der Diskussion auf realwissenschaftlichem Gebiet Ihre Anschauung sehr leicht ins Metaphysische abgleitet. Darüber müsste man sich ausführlich aussprechen. Ich habe einiges darüber an Dr Lutman-Kokoszyńska geschrieben.

Wenn sie meinen, dass es eine Trivialität ist zu sagen, man spreche mit der Sprache über die Sprache, so kann ich darauf nur sagen, dass die Wissenschaft zu einem wesentlichen Teil darin besteht Trivialitäten gegen Irrtümer zu vertreten. Ich habe z.B. vom Beginn des Wiener Kreises an mich gegen die Versuche von WITTGENSTEIN gewehrt eine Art "Erläuterungen" also "nichtlegitime", quasi nicht- oder vorsprachliche Betrachtungen einzuführen, um dann über

Tarski replied:

> I completely agree that "to defend trivialities against errors" is an important task of science. I have indeed for this very reason stressed many times that one must always speak in a language about another language – and not outside the language (from the reductive standpoint, just about my entire 'semantics' should be seen as a triviality; this does not upset me in the least). It seems to me that it is a big mistake, when Wittgenstein, Schlick etc. speak of "the" language instead of (a number of) languages; that might be the true source of the Wittgensteinian "metaphysics". Incidentally, all those who speak about the unified language of science with the slogan "Unity of Science" [Einheitswissenschaft] seem to commit the same mistake. We all know – because of arguments from semantics and syntax– that there is strictly speaking no unified language [Einheitssprache] in which science as a whole could be expressed. It is not enough to say that this is just a temporary, imprecise formulation. For, what should then the final, precise formulation be? Kokoszyńska recently held a lecture on the problem of a Unified Science for the local philosophical society and subjected this point to her criticism; an article from her on this subject is forthcoming in Polish.
>
> (Tarski to Neurath, 28.IV.36)[8]

Kokoszyńska's article appeared in *Erkenntnis* 1937–1938. Tarski's point, as can be gathered from other sources, is that a unified language for science requires a considerable amount of mathematics. And thus, on account of Gödel's incompleteness theorems, any such language for science will be incomplete. Then Tarski moves on to talk about the aims and background for his definition of truth.

> Now, as far as my "terminological choice" is concerned, I can assure you, firstly, that it came about completely independently of Wittgenstein's metaphysics and, secondly, that

---

die Gegenüberstellung von "der" Sprache mit "der" Wirklichkeit zu reden, also *ausserhalb* der Sprache. Ich glaube, dass die "Konstatierungen" von Schlick, die Sätze und doch wieder nicht Sätze sind aus dieser WITTGENSTEINSCHEN Metaphysik herzuleiten sind.

Und sofern Ihre terminologische Wahl bedenkliche Konsequenzen nahelegt, ist sie vielleicht nicht ganz unabhängig von diesen Konsequenzen zustandegekommen. Auf der einen Seite wird betont, dass dieser Wahrheitsbegriff nur für formalisierten Sprachen gelte, andererseits ist der Wahrheitsbegriff gerade in nicht formalisierten Bereich von praktischer Bedeutung. Deshalb bin ich, wenn man den Terminus nicht überhaupt fallen lässt mehr für meine Terminologie, die im nicht formalisierten Bereich verwendbar bleibt. Während die von Ihnen und Lutman verwendete Terminologie im nicht formalisierten Bereich verwendet zu schlimmen Dingen führt. (Neurath to Tarski, 24.IV.36, Neurath Nachlaß)

[8] Ich bin völlig Ihrer Meinung, daß es eine wichtige Aufgabe der Wissenschaft ist "Trivialitäten gegen Irrtümer zu vertreten". Eben deshalb habe ich ja selbst vielmals betont, daß man stets in einer Sprache über eine andere Sprache sprechen muß – und nicht außerhalb der Sprache (vom rein deduktiven Standpunkte aus ist übrigens meine ganze "Semantik" fast als eine Trivialität anzusehen; das ärgert mich nicht im wenigsten). Es ist – wie mir scheint – ein großer Fehler, wenn Wittgenstein, Schlick usw. von "der" Sprache anstatt von Sprachen (in Mehrzahl) sprechen; das ist vielleicht die echte Quelle der Wittgensteinschen "Metaphysik". Nebenbei gesagt, denselben Fehler scheinen auch alle diejenigen zu begehen, die im Zusammenhang mit dem Stichwort "Einsheitswissenschaft" über die Einheitssprache der Wissenschaft reden. Wir wissen ja alle – auf Grund der Erörterungen aus der Semantik und Syntax –, daß es streng genommen keine Einheitssprache gibt, in der die ganze Wissensschaft ausdrückbar wäre. Es genügt nicht zu sagen, daß das nur eine vorläufige, unpräzise Formulierung ist; denn wie soll die endgültige, präzise Formulierung lauten? (Kokoszyńska hatte vor kurzer Zeit einen Vortrag über das Problem der Einheitswissenschaft in der hiesigen Phil. Gesell. und hat u.a. diesen Punkt einer Kritik unterworfen; es soll ein Aufsatz von ihr in der polnischen Sprache darüber erscheinen). (Tarski to Neurath, 28.IV.36, Neurath Nachlaß)

> it was in no way a "choice". The problem of truth came up very often, especially in the Polish philosophical literature. One was constantly asking (see for instance Kotarbiński's "Elements"), whether it was possible to define and apply the concept of truth unobjectionably, using such and such properties (which I spelled out in my later work). I simply provided a positive solution to this problem and noted that this solution can be extended to other semantic concepts. Like you, I am certain that this will be misused, that a number of philosophers will "overinterpret" this purely logical result in an unacceptable manner. Such is the common destiny of both small and great discoveries in the domain of the exact sciences (at times, one compares the philosophers to the "hyenas of the battle field").
>
> (Tarski to Neurath, 28.IV.36)[9]

He also comments on his position vis-à-vis metaphysics that marks a stark contrast to Neurath's iconoclastic attitude:

> But I must confess to you that even if I do not underestimate your battle against metaphysics (still more from a social than from a scientific point of view), I personally do not live in a constant and panic fear of metaphysics. As I recall, Menger once wrote something witty on the fear of antinomies; it seems to me that one could apply it—mutatis mutandis—to the fear of metaphysics. It is a hopeless task to caution oneself constantly against metaphysics. This becomes all the clearer to me when I hear, here at home, various attacks on the very metaphysics of the Vienna Circle (going, namely, in your direction and in that of Carnap), when, for instance, Łukasiewicz talks, with respect to the "Logical Syntax", about Carnap's philosophy, philosophizing etc. (in his mouth this has roughly the same sense as 'metaphysics' in yours). What you blame me for on account of the concept of truth, one blames Carnap for on account of the introduction of the terms 'analytic', 'synthetic', etc. (Regression to the Kantian metaphysics); and it seems to me that I was even more justified than Carnap to designate as truth the concept that I discuss. In general it is a valuable task to fill old bottles with new wine.
>
> (Tarski to Neurath, 28.IV.36)[10]

---

[9] Was nun meine "terminologische Wahl" betrifft, so kann ich Ihnen versichern, daß sie erstens ganz unabhängig von der Wittgensteinschen Metaphysik zustandegekommen ist und daß es zweitens überhaupt keine "Wahl" war. Das Problem der Wahrheit kam speziell in der polnischen Philosophischen Litteratur sehr oft vor, man hat immer gefragt, ob man den Wahrheitsbegriff mit den und den Eigenschaften (die ich später in meiner Arbeit genau präzisiert habe) in einwandfreier Weise definieren und verwenden kann (vgl. Z.B. die "Elemente" von Kotarbiński). Ich habe einfach dieses Problem positiv gelöst und habe bemerkt daß sich diese Lösung auf andere semantische Begriffe ausdehnen läßt. Ebenso wie Sie bin ich sicher, daß man daraus verschieden Mißbräuche machen wird, daß verschiedene Philosophen dieses Ergebnis rein logischer Natur in unerläßlicher Weise "hinausinterpretieren" werden – das ist das gemeinsame Schicksal aller kleineren und größeren Entdeckungen aus dem Bereiche der exakten Wissenschaften (man vergleicht ja manchmal die Philosophen mit den "Hyänen des Schlachtfeldes"). (Tarski to Neurath, 28.IV.36, Neurath Nachlaß)

[10] Aber ich muß Ihnen offen gestehen: wenn ich auch Ihren Kampf gegen die Metaphysik keineswegs unterschätze (noch mehr unter sozialem, als unter wissenschaftlichem Gesichtspunkt), so lebe ich persönlich nicht in einer ständigen, panischen Angst vor der Metaphysik. Wie ich erinnere, hat einmal Menger etwas geistreiches über die Furcht vor Antinomien geschrieben; es scheint mir, daß man das alles -mutatis mutandis- auch auf die Angst vor der Metaphysik übertragen könnte. Es ist eine hoffungslose Aufgabe, sich stets vor dem Vorwurf einer Metaphysik zu warnen. Das wird mir besonders klar, wenn ich hier bei uns verschiedene Angriffe eben auf die Metaphysik des Wiener Kreises (und zwar Ihrer und Carnapschen Richtung) höre, wenn z. B. Łukasiewicz a propos der "Logischen Syntax" über Carnaps Philosophie, Philosophieren usw. spricht (das hat in seinem Mund ungefähr denselben Sinn wie in Ihrem "Metaphysik"). Dasselbe, was Sie mir wegen

In addition, Tarski pointed out that Neurath's critique does not affect solely his work on truth but any formal work in syntax and semantics. Implicit in the comment is the observation that Neurath would thus find himself condemning most of Carnap's work in the *Logical Syntax of Language*:

> Another point in this connection: my concept of truth, you claim, holds only in formalized languages. But on the contrary, the concept of truth is of practical significance precisely in non formalized domains. One can extend this literally to all precise concepts of syntax and semantics (consequence, content, logical and descriptive concepts, etc.): all these concepts can only be related approximately to the non formalized languages (thus to the actual languages of all non formal sciences [Realwissenschaften]): truth here is no exception.
> (Tarski to Neurath, 28.IV.36)[11]

In Mancosu 2008, I reconstruct the nature of Neurath's qualms about semantics through the correspondence with Carnap, Lutman-Kokoszyńska, Hempel, and Tarski. Tarski later described to Popper (letter dated 4.X.37) a meeting with Neurath in The Hague immediately after the 1937 Paris Congress. Tarski claims to have found the right way to deal with Neurath: "hypnotize rather than persuade".[12] The evidence shows that Neurath did not change his mind on this issue. And it befits this volume to quote the following passage by Neurath from a letter (written in English) to Carnap from 1943 in which the danger of semantics is at issue:

> I am really depressed to see here all the Aristotelian metaphysics in full glint and glamour, bewitching my dear friend Carnap through and through. As often, a formalistic drapery and hangings seduce logically-minded people, as you are very much. . ..It is really stimulating to see how the Roman Catholic Scholasticism finds its way into our logical studies, which have been devoted to empiricism.
>
> Scholasticism created Brentanoism, Brentano begot Twardowski, Twardowski begot Kotarbiński, Łukasiewicz (you know his direct relations to the Neo-Scholasticism in Poland),

des Wahrheitsbegriff vorwerfen, wirft man Carnap wegen der Einführung der Termini "analytisch", "synthetisch" u.s.w. vor (Rückkehr zu der Kantschen Metaphysik); und es scheint mir, daß ich im Grunde noch mehr als Carnap berechtigt war den von mir erörterten Begriff als Wahrheit zu bezeichnen. Im allgemeinen ist es eine wertvolle Aufgabe alte Gefässer mit neuem Trunk zu füllen. (Tarski to Neurath, 28.IV.36, Neurath Nachlaß)

11 Noch ein Punkt in diesem Zusammenhang: mein Wahrheitsbegriff gelte nur für die formalisierten Sprachen, andrerseits ist der Wahrheitsbegriff gerade im nicht formalisierten Bereich von praktischer Bedeutung. Das kann man wörtlich auf alle präzisen Begriffe der Syntax und Semantik (Konsequenz, Gehalt, logischer und deskriptiver Begriff u.s.w.) übertragen: alle diese Begriffe können nur annährungsweise auf die nicht-formalisierten Sprachen (also auf die aktuellen Sprachen aller Realwissenschaften) bezogen werden; Wahrheit ist hier keine Ausnahme. (Tarski to Neurath, 28.IV.36, Neurath Nachlaß)

12 Mit Neurath habe ich noch später in Haag über die Wahrheit gesprochen und habe, wie mir scheint, den richtigen Weg gefunden: hypnotisieren anstatt überzeugen. Er beginnt mir schließlich zu glauben, daß der Wahrheitsbegriff, wenn auch nicht ein besonders wichtiger, so doch ein ganz harmloser, korrekter und ungefährlicher (vom anti-metaphysischen Standpunkt aus) Begriff ist, daß die semantische Auffassung aus diesem Begriff den metaphysischen Stachel mit allen Wurzeln ausreißt, daß im Gegenteil – sein (d.i. Neuraths) Kampf gegen der Wahrheitsbegriff eine eigenartige "Metaphysik à rebours" darstellt. Komisch, nicht wahr? (Tarski to Popper, 4.X.37, Popper Archive)

both together begot now Tarski etc., and now they are God fathers of OUR Carnap too; in this way Thomas Aquinas enters from another door Chicago. (January 15, 1943, RC 102-55-02).

9.3 Tarski on Reducibility in Physicalism

The discussion on truth within the Vienna circle had, up to 1935, been carried out within the larger context of the debate on protocols.[13] It is the acceptance of Tarski's theory of truth that allows Carnap to make a clear distinction between truth and confirmation (Carnap 1936a). This move was made in 1935 in Paris. Meanwhile, Carnap was emphasizing more and more the importance of keeping the logical and psychological aspects of epistemology distinct. Carnap's 'Testability and Meaning' (1936b) draws on these distinctions and puts him on a collision path with Neurath's understanding of what a logic of science is supposed to achieve. Michael Friedman has characterized the situation as follows:

> Here [in 'Testability and Meaning'] the fundamental tension between Carnap's conception of Wissenschaftslogik and Neurath's has become intolerable. According to Neurath's naturalistic understanding of this discipline, there is only the single unified language of empirical science. There is no room, therefore, for a metalanguage or syntax language describing the process of empirical testing from some idealized point of view outside the language of empirical science itself.
>
> (Friedman 2003, 102)

One of the reasons Neurath adduces against the type of logical analysis proposed by Carnap is that the concepts and sentences of the languages under investigation are usually imprecise and fluid [schmierige Ballungen]. It is in this context of opposition between the Carnapian conception of the logic of science and Neurath's naturalism that Hempel writes to Neurath a letter in which he reports a conversation with Tarski who provides a powerful objection to the Carnapian project. To set the context I will also provide the preceding letter from Neurath to Hempel reporting on Tarski's visit. On September 19, 1938, Neurath wrote to Hempel:

> Tarski was here yesterday. As usual very stimulating—but he takes a surprisingly dismissive attitude vis-à-vis the most recent developments of the Vienna Circle. Whilst we think that we want to push sciences to the foreground and do not want to promote "thrashing out", so to speak, as a separate activity (see Waismann's preface to Schlick's shorter writings), he thinks that Wittgenstein's was still genuine anti-philosophy etc. He claimed that Carnap (whose qualities he nonetheless fully recognizes) is, like us all, on the wrong path and that the Encyclopedia is interesting insofar as it brings interesting articles. On the other hand, he says he does not value our effort to give a certain [...] coherence to the articles [...] I tried to make clear to him what we mean by our program of a Unified Science and how little this program is affected by Kokoszyńska's criticisms (the same holds for Popper in Copenhagen). To which he said that he could only hold on to the *wording* in Carnap, me and others. He said he made an effort to give a correct formulation to what we may be meaning but that he did not succeed. [...] We were together a lot but his "mood" vis-à-vis us oppressed me [...] I do not have the impression that he cares about empiricism. For he

[13] For a thorough analysis of the debate on protocols in the Vienna Circle see Uebel 1992.

> broached the question of TRUTH by saying only this: obviously, thinkers such as Scholz, Kotarbiński and others are now in the position to use a concept which once seemed to be contradictory, etc. [...] it remains nonetheless (this is my impression) that of all the things the Poles say [...] little speaks in favour of some direct sympathy for our empiricism, but most likely rather for something that harks back to their "ontological" tendencies. I could be mistaken. I have the same feeling about Ajdukiewicz.[14]

By contrast, Hempel's reply described Tarski as seriously thinking about the issues related to empiricism even if in a critical vein:

> Your remarks about the discussion with Tarski interest me very much. On the whole I had a very pleasing impression, not only because T[arski] is in general very sharp and stimulating, but also more specifically in reference to the questions of empiricism. Among others, a conversation with him about the logic of testing empirical hypotheses made a very great impression on me. T[arski] thought, of course, that the Wittgensteinian idea of complete verifiability for empirical hypotheses is entirely naïve; but also that, in his opinion, Carnap's logical theory in Test[ability] and Meaning, based on much more liberal principles, did *not* achieve what was desired: in fact he is acquainted with no single example of a reduction-sentence that actually reduces a concept, say of physical theory, to concepts of the observation-language in materially correct fashion (i.e., so that the empirical investigator would agree). All examples known to him, e.g., C[arnap]'s example "soluble", are schematizations, which the empirical [investigator] must view as inappropriate: in fact it can happen that a material is put in water, does not disappear and yet is soluble. And no matter how many additional conditions and clauses one may add, "exceptions" are still always thinkable.
>
> (Friedman's translation, 103–104)[15]

---

[14] "Gestern war Tarski hier. Wie immer sehr anregend – aber er nimmt eine merkwürdig ablehnende Haltung der neueren Entwicklung der Wiener Kreises gegenüber ein, während wir meinen, dass wir die *Sciences*, in der Vordergrund rücken und das "Klären" als sozusagen isolierbare Beschäftigung nicht befördern wollen (siehe Waismann Vorrede zu den kleinen Schriften von Schlick) meint er, dass Wittgenstein noch richtige Antiphilosophie war usw. Carnap (trotzdem er dessen Qualitäten voll anerkennt) sei so wie wir alle auf einem schiefen Wege und die Enzyklopädie soweit interessant, als sie eben wichtige Artikel bringe – hingegen legte er unseren Bemühen eine gewisse (wohl die heute weitergehende) Kohärenz der Artikel zu erzeugen keinen besonderen Wert bei, wie es schien.

Ich suchte ihm klar zu machen, was wir mit Einheitswissenschaft als Programm meinen und wie wenig das durch Kokoszyńska (ähnlich Popper in Kopenhagen) getroffen werde. Worauf er meinte, er könne sich nur an den *Wortlaut* bei Carnap, mir und anderen halten. Er habe sich bemüht den, was wir vielleicht meinen einen korrekten Ausdruck zu verleihen, es sei ihm aber nicht gelungen.

[...] Wir waren sehr mit einander, aber seine "Stimmung" uns gegenüber bedrückte mich. Wie geht das Ihnen? Ich habe so wenig das Gefühl dass er sich um Empirismus bemüht. Denn er erörterte die WAHRHEITsfrage nur so: offenbar sind jetzt Denker, wie Scholz, Kotarbiński und andere in der Lage einen Begriff, der vorher widerspruchsvoll schien zu verwenden usw. Ich will einen Teil seiner eigentlich durchgehend negativen Stimmung auf seine wirklich triste Lage zurückführen (wenn man jemandem helfen sollte, so ihm. Es ist ja greulich so zu leben) aber trotzallem bleibt übrig, dass (meinem Gefühl nach) aus allen, was die Polen sagen [...] wenig unmittelbare Sympathie für unseren Empirismus spricht, wohl aber irgend etwas, was wohl auf die "ontologische" Neigung zurückgeht. Mag sein, dass ich irre. Auch bei Ajdukiewicz habe ich das." Tarski's condition were indeed difficult as can be gathered from a personal report in a letter to Popper dated 3.XII.36. See also Feferman-Feferman 2004.

[15] Ihre Bemerkungen über Besprechung mit Tarski interessieren mich sehr. Ich hatte im ganzen einen erfreulichen Eindruck, nicht nur weil T. überhaupt sehr gescheit und anregend ist, sondern auch mehr im speziellen, was die Fragen des Empirismus angeht. Unter anderm hat mir

Before we look at the consequences Tarski drew from his claim let us say something more about Carnap's theory of reduction. I will follow here Carnap's 'Ueber die Einheitssprache der Wissenschaft. Logische Bemerkungen zum Projekt einer Enzyklopädie' (1936) which treats in a more relaxed fashion the theory expounded in 'Testability and Meaning' (1936). Carnap's treatment of reduction is related to the problem of how to develop a unified science. One of the features of unified science is the rejection of the idea that different branches of sciences use different concepts. For instance, as Carnap wittily puts it, it should be possible to get by with a single unified concept of "cow" as opposed to a multitude of concepts such as "economical cow", "physical cow", "biological cow", etc. This leads to the problem of the introduction of new concepts in the language of unified science on the basis of other concepts that are already given. Carnap begins by discussing two standard procedures for introducing new concepts by definition: explicit and contextual definition. An example of explicit definition might be "x is a bachelor iff x is a male and x is unmarried". An example of definition in context is the definition of the existential quantifier through negation and the universal quantifier. In the latter case any sentence in which an existential quantifier appears can be replaced by an equivalent sentence that does not contain that symbol. In both cases the defined concept can always be eliminated in favor of the other given concepts. However, these two procedures are not sufficient for the needs of a unified language of science. There is a third definitional strategy, which Carnap calls 'reduction'. Carnap uses the example of the introduction of the concept "x is soluble (in water)". Suppose we are already given the two concepts W(x, t) and L(x, t) standing respectively for "x is in water at time t" and "x dissolves itself at time t". A first attempt at a definition of "x is soluble (in water)" might be as follows:

$$\mathrm{Ll}(x) \equiv (t)\,[W(x,t) \supset L(x,t)]$$

ein Gesprach mit ihm über die Logik der Nachprüfung empirischer Hypothesen recht grossen Eindruck gemacht. T. meinte dazu, dass natürlich die Wittgensteinsche Idee der vollständigen Verifizierbarkeit von empirischen Hypothesen durch Beobachtunsätze ganz naiv sei; dass aber s. E. auch Carnaps, auf viel liberaleren Prinzipien fussende, logische Theorie in Test. and Meaning das Gewünschte *nicht* leiste: tatsächlich kenne er kein einziges Beispiel einer Reduktionsformel, welche tatsächlich einen Begriff etwa der physikalischen Theorie auf Begriffe der Beobachtungssprache in inhaltlich korrekter Weise (dh. so, dass der empirische Forscher dem zustimmen wurde) zurückführt. Alle ihm bekannten Beispiele zB. Cs Beispiel "löslich" seien Schematisierungen, welche der Empiriker als unzutreffend ansehen müsse: tatsächlich könne es vorkommen, dass ein Stoff ins Wasser getan werde, nicht verschwinde und doch löslich sei. Und wieviele Zusatzbedingungen und Klauseln man auch Hinzufügen möge: noch immer seien "Ausnahmen" denkbar. Für die Frage der Nachprüfung empirischer Hypothesen habe das folgende Konsequenz: führen wir die in den wissenschaftlichen Hypothesen auftretenden theoretischen Begriffe durch Definitions—oder Reduktionsketten ein, die auf Terme der Beobachtungssprache zurückgehen, so wird jede Hypothese bestimmt falsch, weil es dann eben immer abweichende Fälle gibt. – Bisher liege seines Erachtens keine Theorie vor, die in sachlich adäquater Weise eine logische Verbindung zwischen Theorie einerseits und dem Gebiet der Beobachtungen andererseits herstelle. (ganz ähnliche Ideen hatte interessanterweise Waismann in Cambridge mir gegenüber geäussert).

Unfortunately this won't do since the right hand side of the equivalence is satisfied by any object which is not in water, say a match that has never been in water and has just been burnt. Thus a match, which is not water soluble, would turn out according to this definition to be soluble after all. Carnap therefore proposes, as an accurate rendering of such dispositional terms the following formula which characterizes, according to him, the right relation between the three concepts:

$$(t)[W(x, t) \supset (Ll(x) \equiv L(x, t))]$$

This is called the reduction sentence for Ll on the basis of W and L. Carnap says that through this sentence Ll is reduced to W and L and the meaning of the new concept is determined:

> through this reduction the meaning of the new concept is indeed determined; for we know what we have to do in the individual cases to determine empirically whether the new concept belongs to a certain thing.
>
> (Carnap, 1936b, 64)

That is, one puts the object b in water. If it dissolves then it has the property and if it does not dissolve then it does not have the property. What characterizes this third type of definition in comparison to the previous two is that in general such definitions cannot be eliminated. The sentences in which the new concept occurs are in general not replaceable by sentences in which the concept does not occur. This led Carnap to distinguish between the thesis that every concept of science is definable on the basis of concepts of physical science (now declared false) and the thesis that every concept of science is reducible to concepts of physical science. It is the latter, according to Carnap, that captures the proper version of physicalism.

We can now go back to Tarski. We have already seen that one objection made by Tarski consists in the claim that Carnap has failed to show the reducibility even of his favorite pet concepts such as "x is soluble (in water)". The reason Tarski gave is that "in fact it can happen that a material is put in water, does not disappear and yet is soluble. And no matter how many additional conditions and clauses one may add, 'exceptions' are still always thinkable." Michael Friedman has drawn attention to these comments by Tarski. He was especially struck by the fact that Hempel himself was to use exactly Tarski's argument in his paper 'Provisoes' (1988) in which he explicitly repudiated the Carnapian style of analysis and moved to a model closer to Neurath's naturalism. As Friedman puts it:

> What I find most remarkable here is that it is essentially Tarski's argument, although in a clearer and more explicit form, that constitutes the centerpiece of Hempel's 'Provisoes' (1988). And this article, published, appropriately enough, in *Erkenntnis*, is in turn the centerpiece of Hempel's later conversion, noted several times above, from the Carnapian program of logical "explication" or "logical reconstruction" to a more naturalistic emphasis on historical, sociological, and other broadly "pragmatic" factors.
>
> (Friedman, 2003, 104)

I should only add that Carnap had also pointed out the problem with reduction sentences in Carnap 1956. There are other parts of the letter that were not quoted in Friedman's article (because his aim was more to point out the Tarskian roots of some of Hempel's views as opposed to focusing on Tarski). Tarski, as reported by Hempel, drew the following conclusion from his thesis:

> If we introduce the theoretical concepts appearing in scientific hypotheses through definitions, or chains of reductions, which go back to terms of the observation language, then every hypothesis will turn out to be certainly false because there are always disconfirming instances. In his opinion we so far have no theory that erects in materially adequate fashion a logical connection between theory, on the one side, and the realm of observations, on the other (interestingly, Waismann expressed to me quite similar ideas in Cambridge).
>
> Tarski's hesitations are shared (and are perhaps in part stimulated) by Wundheiler and Poznański. However, whereas the latter, as T[arski] indicated, are still hopeful about the search for a logical bridge (as Carnap and, e.g., I, in his opinion, still take one to be constructible), it appears to him that it is at the very least not excluded that no such bridge can be forged in an adequate manner, and that the evaluation of theories by means of observations occurs perhaps instinctively, as it were, without it being the case that predictions be first deduced theoretically in the form of observation-sentences with the help of reduction-sentences, etc.[16]

Thus the consequence Tarski drew from the inadequacy of the reduction sentence was that in such a reconstructed physicalist system of science every single hypothesis of science would turn out to be false. Finally, in the final part of the letter Hempel reports some other thoughts of Tarski that give us a glimpse of Tarski's thinking at the time:

> Tarski even thinks that it might be better to consider the domain of observations (e.g. "experimental physics") as a system of operations and to avoid formulating the experimental results through sentences that are then confronted with the theory. On the other hand he thinks it possible that something like a many valued logic could lead to a satisfactory formalization of the relation between theory and observation. While in our logic a single deviant observation makes a universal claim absolutely false, in such a system of many valued logic the finding of a negative instance would only yield a smaller truth-value

---

[16] "Führen wir die in den wissenschaftlichen Hypothesen auftretenden theoretischen Begriffe durch Definitions- oder Reduktionsketten ein, die auf Terme der Beobachtungssprache zurückgehen, so wird jede Hypothese bestimmt falsch, weil es dann eben immer abweichende Fälle gibt. Bisher liege seines Erachtens keine Theorie vor, die in sachlich adäquater Weise eine logische Verbindung zwischen Theorie einerseits und dem Gebiet der Beobachtungen andererseits herstelle. (Ganz ähnliche Ideen hatte interessanterweise Waismann in Cambridge mir gegenüber geäussert.) Tarskis Bedenken werden geteilt (und sind vielleicht zT. angeregt) von Wundheiler und Poznański. Während diese aber, wie T. andeutete, doch hoffungsvoll auf der Suche nach einer logischen Brücke seien (wie sie Carnap und zB. ich s.E. doch auch für konstruierbar hielten) scheint es ihm wenigstens nicht ausgeschlossen, dass keine solchen Brücken in adäquater Weise geschlagen werden können, und dass die Beurteilung von Theorien an Hand von Beobachtungen vielleicht sozusagen instinktiv erfolge, ohne dass erst Prognosen in Form von Beobachtungssätzen mit Hilfe der Reduktionsformeln theoretisch deduziert werden, usf." Wundheiler and Poznański 1934 contains important material which however was accessible only partially to Western scholars (as it was written in Polish). Carnap, for instance, made use of a partial translation into German made by Rose Rand. The translation is preserved in the Carnap Nachlaß under RC 081-37-01.

for the hypothesis than the one previously given to it. (This conjecture points in a direction which is the one Reichenbach and now also Carnap turn to in order to look for the "bridge").[17]

Jan Woleński has pointed out to me that Poznański and Wundheiler were influenced by Bridgman's operationalism. The above comments are obviously connected to such a position although more work would be needed to say something more precise about a possible (direct or indirect) influence of Bridgman on Tarski.

9.4 Nominalistic Construction of Mathematics and Science

In 1942, Quine wrote to Woodger:

> Last year logic throve. Carnap, Tarski and I had many vigorous sessions together, joined also, in the first semester, by Russell. Mostly it was a matter of Tarski and me against Carnap, to this effect. (a) C[arnap]'s professedly fundamental cleavage between the analytic and the synthetic is an empty phrase (cf. my 'Truth by convention'), and (b) consequently the concepts of logic and mathematics are as deserving of an empiricist or positivistic critique as are those of physics. In particular, one cannot admit predicate variables (or class variables) primitively without committing oneself, insofar to the "reality of universals", for better or worse; and meanwhile C.'s disavowal of "Platonism" is an empty phrase (cf. my 'Description and Existence'). Other points on which we took C. to task are (c) his attempt to make a general semantics rather than sticking to a convenient canonical form for object languages and studying the semantics thereof more simply and briefly and yet more in detail; (d) his resuscitation of intensional functions. C. argued reasonably and well, as always, and the discussions were good fun.
>
> (Quine to Woodger, May 2, 1942, Woodger papers)

The sessions in question were those of the year 1940–1941, which took place at Harvard and to which both Quine and Carnap devote much space in their autobiographies. For instance, Carnap writes:

> During the year 1940–1941 I was a visiting professor at Harvard. During the first semester Russell was there too, giving the William James lectures, and I was glad to have an even better opportunity for talks with him on questions of philosophy as well as on social and political issues. Tarski spent the same year at Harvard. We formed a group for the discussion of logical problems; Russell, Tarski, Quine and I were its most active members. I gave several talks on the nature of logic and on the possibility of defining logical truth

---

[17] T[arski] meint sogar, dass *vielleicht* es besser sein mag, das Gebiet der Beobachtungen (zB. die "Experimentalphysik") als ein System von Handlungen zu betrachten, und die Versuchsresultate nicht durch Sätze zu formulieren, die dann mit der Theorie konfrontiert werden. Andererseits hält er es für denkbar, dass irgend so etwas wie eine mehrwertige Logik zu einer befriedigenden Formalisierung der Beziehungen zwischen Theorie und Beobachtung führen könne: Während in unserer Logik eine einzige abweichende Beobachtung ein universelles Gesetz unerbittlich falsch mache, würde in einem solchen mehrwertigen System die Hypothese bei Auffindung einer negativen Instanz nur einen kleineren Wahrheitswert erhalten als zuvor. (Diese Vermutung weist in eine richtung, in der Reichenbach und jetzt auch Carnap die "Brücke" suchen).

> as a semantical concept. I discovered that in these questions, even though my thinking on semantics had originally started from Tarski's ideas, a clear discrepancy existed between my position and that of Tarski and Quine, who rejected the sharp distinction I wished to make between logical and factual truth.
>
> (Carnap 1963, 35–36; see also 64–65)

Moreover, there was substantial discussion on other issues:

> In other problems we came to a closer agreement. I had many private conversations with Tarski and Quine, most of them on the construction of a language of science on a finitistic basis.
>
> (Carnap 1963, 36)

Luckily, Carnap took notes during these meetings. They are still found in his archive in Pittsburgh and make up more than 80 pages of typescript transcription. They are invaluable for us, as Carnap reports at length on Tarski's position on issues such as the distinction between the analytic and the synthetic, Tarski's views on languages and on his nominalism. Once again, I have a paper on these discussions (Mancosu 2005) and here I will simply discuss the issues related to the analytic/synthetic distinction and to nominalism by providing some quotation to give an idea of the material.

9.4.1 Tarski on the Carnapian Distinction Between Analytic and Synthetic

We have seen from the citations given above that Tarski objected to a sharp distinction between analytic and synthetic sentences (L-truth, F-truth). Quine had the same misgivings, which in his letter to Woodger he claims as being present already in 1936, but in the case of Tarski we can show that they go back to the beginning of the thirties.

Haller quotes a note in Carnap's diary, dated February 22, 1930 that reads:

> 8–11 with Tarski at a Café. About monomorphism, tautology, he will not grant that it says nothing about the world; he claims that between tautological and empirical statements there is only a mere gradual and subjective distinction.
>
> (quoted in Haller 1992, 5)[18]

Tarski reasserted his position on several occasions. For instance, in a report entitled 'Conversation with Tarski' Carnap reports the following:

> 'L-true'. 'logic-descriptive'. I [Carnap]: my intuition, Clearer in the distinction L-true-F-true, than in logic-descriptive. But I can always explain the latter by displaying the simplest logical constants in the usual systems and by claiming that everything that can be defined on their basis should be logical. He [Tarski]: he has no such intuition; one could just as well

[18] 8–11 h mit Tarski im Café. Über Monomorphie, über Tautologie, er will nicht zugeben, daß sie nichts über die Welt sagt; er meint zwischen tautologischen und empirischen Sätzen sei ein bloß gradueller und subjektiver Unterschied. (quoted in Haller 1992, 5)

> consider [the sign] "temperature" to be logical. (RC 090-16-09 Dated 6. 3.40, Chicago, Conversation with Tarski [4pp.])[19]

We have seen that Quine in his letter to Woodger claimed that he and Tarski had argued against Carnap that "the concepts of logic and mathematics are as deserving of an empiricist or positivistic critique as are those of physics." This, as Quine pointed out, was a consequence of rejecting the analytic/synthetic distinction. One of the consequences, in particular, was the idea that logic and mathematics are just as revisable as physics. A glimpse of this position can already be seen from a comment Tarski made during a meeting of the Vienna circle in June 1935. In a letter from Neider to Neurath, dated 29.6.35, Neider gave a summary of a discussion on protocols in the Circle and quotes Tarski as claiming: "I have never uttered a sentence whose revisibility I excluded."[20] A very clear statement of this position can be found in Tarski's letter to Morton White dated September 23, 1944:

> I think that I am ready to reject certain logical premisses (axioms) of our science [logic] in exactly the same circumstances in which I am ready to reject empirical premisses (e.g., empirical hypotheses); and I do not think that I am an exception in this respect...I can imagine that certain new experiences of a very fundamental nature may make us inclined to change just some axioms of logic. And certain new developments in quantum mechanics seem clearly to indicate this possibility. That we are reluctant to do so is beyond any doubts; after all, 'logical truths' are not only more general, but also much older than physical theories or even geometrical axioms. And perhaps we single out these logically true sentences, combine them in a class, just to express our reluctance to reject them.
>
> (Tarski 1944, 31–32)

Morton White, in his preface to the publication of Tarski's letter we have just quoted, points out two aspects of Tarski's thought that influenced him and Quine. The first is the idea that logic and mathematics might be just as revisable as empirical science and the second is the rejection of the analytic/synthetic distinction. The passages quoted above show that White's claim is supported by the historical documents.

Moreover, the documents show that the analytic/synthetic distinction (in the form dependent from Wittgenstein's formulation opposing tautological and empirical statements) had been criticized by Tarski already in 1930. It is quite likely that the idea that logic and mathematics might be revisable dates from the same period, although the first explicit statement I was able to find dates from the report given by Neider in 1935.

It is unfortunate that the Carnap notes on the 1940–1941 meetings do not have much to contribute to this criticism of the analytic/synthetic distinction on the part

[19] 1. 'L-wahr'. 'logisch-deskri[ptiv]'. Ich: meine Intuition. Klarer in der Unterscheidung L-wahr—F-wahr, als in logisch-descri[ptiv]. Die Letztere kann ich aber immerhin erklären durch Aufweisung der einfachsten logischen Konstanten in den üblichen Systemen und Angabe, dass Alles daraus Definierbaren auch logisch sein soll. Er: Er hat gar keine solche Intuition; man könnte ebenso gut 'Temperatur' auch als logisch rechnen. (RC 090-16-09 Dated 6. 3.40, Chicago, Conversation with Tarski [4 pp.])

[20] "Tarski: Ich habe noch nie einen Satz gesprochen, dessen Korrigierbarkeit ich ausgeschlossen habe", Letter from Neider to Neurath, dated 29.6.35, Neurath Nachlaß.

of Quine and Tarski. By contrast, they are extremely informative on other topics. In the next section I will briefly treat the topic of nominalism. For a more extended treatment of these discussions see Mancosu 2005.

9.4.2 Nominalism

As we have seen at the outset, it is known that Tarski had nominalistic sympathies but he was not forthcoming concerning the exact nature of his nominalism. Nominalism was certainly important to him. In a letter to Woodger, written in 1948, he wrote:

> The problem of constructing nominalistic logic and mathematics has intensively interested me for many-many years. Mathematics – at least the so-called classical mathematics—is at present an indispensable tool for scientific research in empirical science. The main problem for me is whether this tool can be interpreted or constructed nominalistically or replaced by another nominalistic tool which should be adequate for the same purposes
> (Tarski to Woodger, November 21, 1948, Woodger Papers)

The Carnap transcripts of the 1940–1941 meetings are the best source for giving us a more detailed picture of Tarski's position on nominalism. A number of nominalistic tenets appear already in the report of discussions concerning the nature of typed vs untyped languages. Consider the following conversation:

> I [Carnap]: Should we construct the language of science with or without types?
> He [Tarski]: Something entirely different might emerge. One would hope and perhaps conjecture that the entire general set theory, however beautiful it is, will in the future disappear. Platonism begins with the higher types. The tendencies of Chwistek and others (Nominalism) to speak only of what can be named are healthy. The problem is only to find a good execution.[21]

Tarski comes back to the same claim about the Platonist commitment involved in higher order quantifications several times during the 1940–1941 meetings. For instance:

> Tarski: A Platonism underlies the higher functional calculus (thus the use of a predicate variable, especially of higher type) (102-63-09)[22]

It turns out that the project of a nominalistic foundation for mathematics and science was at the center of the discussions between Tarski, Quine and Carnap in 1941. Here is how Carnap summarizes those discussions:

---

[21] Ich: Sollen wir vielleicht die Sprache der Wissenschaften mit oder ohne Typen machen? Er: Vielleicht wird sich etwas ganz Anderes entwickeln. Es wäre zu wünschen und vielleicht zu vermuten, dass die ganze allgemeine Mengenlehre, so schön sie auch ist, in der Zukunft verschwinden wird. Mit den höheren Stufen fängt der Platonismus an. Die Tendenzen von Chwistek und anderen (Nominalismus), nur über Bezeichenbarem zu sprechen, sind gesund. Problem nur, wie gute Durchführung zu finden. (RC 090-16-09)

[22] Tarski: Ein *Platonismus* unterliegt dem höheren Funktionskalkul (also den Gebrauch einer Prädikatenvariable, besonders höherer Stufe). (RC102-63-09)

My thinking on these problems received fruitful stimulation from a series of conversations which I had with Tarski and Quine during the academic year 1940–1941, when I was at Harvard; later Nelson Goodman participated in these talks. We considered especially the question which form the basic language, i.e. the observation language, must have in order to fulfill the requirement of complete understandability. We agreed that the language must be nominalistic, i.e., its terms must not refer to abstract entities but only to observable objects or events. Nevertheless, we wanted this language to contain at least an elementary form of arithmetic. To reconcile arithmetic with the nominalistic requirement, we considered among others the method of representing the natural numbers by the observable objects themselves which were supposed to be ordered in a sequence; thus no abstract entities would be involved. We further agreed that for the basic language the requirements of finitism and constructivism should be fulfilled in some sense. We examined various forms of finitism. Quine preferred a very strict form; the number of objects was assumed to be finite and consequently the numbers occurring in arithmetic could not exceed a certain maximum number. Tarski and I preferred a weaker form of finitism, which left it open whether the number of all objects is finite or infinite. Tarski contributed important ideas on the possible forms of finitistic arithmetic.

(Carnap 1963, 79)

The notes found in Carnap *Nachlass* give us a more detailed view of such discussions and provide us with much more information on the nature of Tarski's nominalism. For instance, on (10.1.41), Tarski describes as follows his nominalistic commitments:

I understand basically only languages which satisfy the following conditions:

1. Finite number of individuals
2. Realistic [reistic? – PM] (Kotarbiński): the individuals are physical things;
3. Non-platonic: it includes only variables for individuals (things) not for universals (classes and so on)

Other languages I "understand" only the way I "understand" [classical] mathematics, namely as a calculus; I know what I can infer from other [sentences] (or have inferred; "derivability" in general is itself problematic). In a discussion, I always interpret higher "Platonic" assertions [Aussagen] as asserting that a given proposition is derivable (or derived) from others. (He most likely means the following: the assertion of a given proposition is interpreted as saying: this proposition holds in the determinate system that is presupposed; and this means: it is derivable from given basic assumptions).

Why is elementary arithmetic, with a countable domain, already excluded? Because, according to Skolem, all of classical mathematics can be represented through a countable model, and can therefore be expressed in elementary arithmetic, by taking 'e' as a specific relation between natural numbers.[23]

---

[23] *Tarski, Finitismus.* Bemerkung in Diskussion in der Logikgruppe, *10.1.41.* Tarski: Ich verstehe im Grunde nur eine Sprache die folgende Bedingungen erfüllt: [1] *Finite* Anzahl der Individuen; [2] *Realistisch* (Kotarbiński): Die Individuen sind physikalische Dinge; [3] Nicht-platonisch: Es kommen nur Variable für Individuen (Dinge) vor, nicht für Universalien (Klassen usw.). Eine andere Sprache "verstehe" ich nur so, wie ich die klassische Mathematik "verstehe", nämlich als Kalkül; ich weiss, was ich aus anderem Ableiten kann (oder abgeleitet habe; "Ableitbarkeit" im Allgemeinen schon problematisch). Bei irgendwelchen höheren, "platonischen" Aussagen in einer Diskussion deute ich sie mir als Aussagen, dass ein bestimmter Satz aus gewissen anderen Sätzen ableitbar (bzw. abgeleitet) ist. (Er meint wohl so: Die Behauptung eines gewissen Satzes wird

The particular approach to nominalism, dictated by the idea that we should rely on non-abstract concrete objects, led to quite a tension between Tarski and Quine on one side and Carnap on the other side. Here is a passage concerning reflections that Carnap writes for himself on the difference between his view and that of Tarski:

> Tarski's finitism is a logical one; he thinks: the number of things in the world is perhaps finite; in this case one can only speak of finitely many natural numbers. By contrast I [Carnap] say: we are empiricists. Therefore we say: our knowledge is limited to the finite; that is, each confirmation is based on a finite set of evidence, that is, a finite set of observational expressions. But: we can nonetheless speak about finite classes of arbitrarily high cardinality, thus also about the individual natural numbers (for instance, 1000≠1001), without taking into consideration the number of things in the world. Thus, *logic and arithmetic become independent of the contingent number of things in the world*. Nonetheless, if they have to be really understood, logic and arithmetic remain, in a certain other sense, finitistic. The arithmetic (of the natural numbers) has in fact been developed without our knowing (up to this day) with certainty whether the number of things in the world is finite or not. And the demonstrated propositions are not doubted by anyone; the concrete propositions (i.e. without variables), in particular, do appear indubitable. Thus arithmetic can indeed be independent of a factual hypothesis about the world. Even if the number of things (for instance, electrons etc.) is finite, the number of events can nonetheless be assumed to be infinite (not only the number of moments within an interval on account of density, but also the number of moments which lie one unit from one another, in other words: infinite length of time). Is this a factual hypothesis? Or is this not also in turn related to logical possibility?[24]

---

gedeutet als besagend: dieser Satz gilt in dem bestimmten, vorausgestzten System; und das heisst: er ist ableitbar aus gewissen Grundannahmen). Warum wird auch schon die elementare Arithmetik, mit abzählbarem Bereich, ausgeschlossen? Weil, nach Skolem, die ganze klassische Mathematik sich durch ein abzählbares Modell darstellen lässt, also in der elementare Arithmetik ausdrücken lässt, z.B. indem man ε als eine gewisse Beziehung zwischen natürlichen Zahlen nimmt. (RC 090-16-28)

24 *Empiristischer vs. logischer Finitismus.* Tarskis Finitismus ist ein logischer. Er meint: vielleicht die Anzahl der Dinge in der Welt endlich; in diesem Fall kann man auch nur von endlich vielen natürlichen Zahlen sprechen, Ich dagegen: Wir sind Empiristen. Daher sagen wir: unser Wissen ist auf Endliches beschränkt; d.h. jede Konfirmation ist basiert auf eine endliche Menge von Evidenz, d.h. endliche Menge von Beobachtungsausdrücke. *Aber*: Wir können trotzdem über endliche Klassen von beliebig höher Kardinalzahl sprechen, also auch über die einzelnen natürlichen Zahlen (z.B. 1000≠1001), ohne die Anzahl der Dinge in der Welt inbetracht [*sic*] zu ziehen. So werden *Logik und Arithmetik unabhängig von der zufälligen Anzahl der Dinge in der Welt.* Trotzdem bleiben auch Logik und Arithmetik in einem gewissen anderen Sinn finitistisch, wenn sie wirklich verstanden werden soll. Die Arithmetik (der natürlichen Zahlen) ist ja tatsächlich entwickelt worden, ohne dass wir [bis heute] mit Sicherheit wissen, ob die Anzahl der Dinge in der Welt endlich ist oder nicht. Und die bewiesenen Sätze werden von niemandem bezweifelt; besonders die konkreten Sätze (d.h. ohne Variable) scheinen doch unzweifelhaft. Also kann die Arithmetik doch wohl abhängig sein von einer faktischen Hypothese über die Welt. Auch wenn die Anzahl der Dinge (z.B. Elektronen usw) endlich ist, so kann trotzdem die *Anzahl der Ereignisse als unendlich* [angenommen] werden (nicht nur die Anzahl der Zeitpunkte innerhalb eines Intervalls infolge der Dichte, sondern auch die Anzahl der Zeitpunkte im Einheitsabstand von einander, mit anderen Worten: unendliche Länge der Zeit). Ist dies eine faktische Hypothese? Oder hängt es nicht auch wieder mit *logischer Möglichkeit* zusammen? (RC 090-16-24)

It should be obvious to the reader that while Carnap speaks about finitism, this is different from Hilbert's finitism. I will not spell out here how Tarski suggested to proceed by considering only the actually inscribed expressions, sentences, and proofs. I cover this material in detail in Mancosu 2005 and thus I will conclude with a quotation by Goodman-Quine 1947 which shows the essential role Tarski played in motivating work on nominalist foundations of mathematics and science:

> the idea of dealing with the language of classical mathematics in terms of a nuclear syntax language that would meet nominalistic demands was suggested in 1940 by Tarski. In the course of that year the project was discussed among Tarski, Carnap, and the present writers, but solutions were not found at that time for the technical problems involved
> (Goodman, Quine, 1947,112, note 12)

9.5 Conclusion

I realize that, within the limits of this paper, I was not able to give the full context for interpreting the Tarskian quotations I gave nor was I able fully to convey the extent and significance of the Tarskian contribution. For this reason, I said at the beginning that I would proceed impressionistically. I hope the evidence provided has at least convinced the reader that there is much to be found out about Tarski's philosophical and scientific activity by exploring the unpublished resources available to us. More work of this sort will certainly allow us to gauge more precisely Tarski's philosophical positions on a variety of philosophical issues. However, I want to point out that discovering, say, what Tarski's position was on the foundations of mathematics is not the major drive of my research. I am less interested in Tarski's opinion per se as much as in the role Tarski played in the philosophical context surrounding him. And I think the beauty of the material we are uncovering is exactly that of discovering Tarski's pervasive and far-reaching influence in some of the major philosophical discussions of his time.

Acknowledgments I would like to thank S. Feferman, M. Giaquinto, J. Hafner, and P. Wagner for many useful comments on a previous draft of this paper. I am also grateful to Sandra Lapointe for improving the style of the translations from German into English. Many thanks also to the audience and the organizers of the meeting "Le rayonnement de la philosophie polonaise au XXème siècle" (ENS, Paris, 26-7 November 2004) for which occasion the paper was written. Many thanks also to the Special Collections, University of Pittsburgh, and the Wiener Kreis Stiching (Amsterdam) for permission to quote from the Nachlässe of Carnap and Neurath.

References

Archival documents

From Neurath's archive (Haarlem, Konstanz):

Carnap-Neurath correspondence.
Hempel-Neurath correspondence.

Kokoszyńska-Neurath correspondence.
Neurath-Tarski correspondence.
Neurath-Neider correspondence.

From Carnap's archive (Pittsburgh, Konstanz):

The Tarski, Carnap, Quine discussions are found in two set of documents. The first set of documents is classified under RC 090-16-02 to 090-16-30 and has the general title "Tarski (and Quine), Finitism, discussion 1941" The second set of documents represents material on Quine and Carnap in 1941: 090-102-63 "Quine disc., 1940-1941".

The correspondence between Carnap and Lutman-Kokoszyńska is classified under RC 088-57. It contains 16 items.

From Tarski's archive (Berkeley):

Tarski, A., Transcript of remarks, ASL meeting, Chicago, Illinois, April 29, 1965, Bancroft Library

From Popper's archive (Hoover Institution, Stanford)

Popper-Tarski correspondence. Box number 354. Folder ID: 8.

From Woodger's archive (University College London, Special Collection, GB 0103 WOODGER):

The Nachlass is still uncatalogued but the correspondence with Quine and Tarski quoted in this paper is in BOX J-Z.

Bibliography

Ayer, Alfred (1977), A part of my life, London, Collins.

Carnap, Rudolf (1934), Logische Syntax der Sprache, Vienna, Springer. Eng. trans. Logical Syntax of Language, London, Routledge and Kegan Paul, 1937.

Carnap, Rudolf (1935), 'Wahrheit und Bewährung', Actes du Congrès International de philosophie scientifique, vol. 4, Paris, Hermann, 18–23.

Carnap, Rudolf (1936a), 'Ueber die Einheitssprache der Wissenschaft. Logische Bemerkungen zum Projekt einer Enzyklopädie', Actes du Congrès International de philosophie scientifique, vol. 2, Paris, Hermann, 60–70.

Carnap, Rudolf (1936b), 'Testability and meaning', Philosophy of Science vol. 3, 419–471, vol. 4, 1–40.

Carnap, Rudolf (1956), 'The methodological character of theoretical concepts', in H. Feigl, M. Scriven (eds.), Minnesota Studies in Philosophy of Science, vol. I, Minneapolis, University of Minnesota Press, 38–76.

Carnap, Rudolf (1963), 'Intellectual Autobiography', in P. A. Schilpp (ed.), The Philosophy of Rudolf Carnap, (Library of Living Philosophers), vol. 11, LaSalle, Open Court.

de Rouilhan, Philippe (1998), 'Tarski et l'universalité de la logique', in F. Nef and D. Vernant (eds.), Le Formalisme en Question. Le Tournant des Années 1930, Paris, Vrin, 85–102.

Feferman, Anita, and Solomon, Feferman (2004), Alfred Tarski, Life and Logic, Cambridge, Cambridge University Press.

Friedman, Michael (2003), 'Hempel and the Vienna Circle', in G. L. Hardcastle and A. W. Richardson (eds.), Logical Positivism in North America, Minneapolis, University of Minnesota Press, 94–114.

Goodman, Nelson, and Willard V. O. Quine (1947), 'Steps towards a constructive nominalism', The Journal of Symbolic Logic 12, 105–122.

Haller, Rudolf (1992), "Alfred Tarski, Drei Briefe an Otto Neurath' Grazer Philosophische Studien 43, 1–32.

Hempel, Gustav (1935), 'On the Logical Positivists' theory of truth' Analysis 2, 49–59; also in Selected Philosophical Essays, R. Jeffrey (ed.), Cambridge, Cambridge University Press, 1999, 21–25.

Hempel, Gustav (1988), 'Provisoes, A problem concerning the inferential function of scientific theories', Erkenntnis 28, 147–164; also in Selected Philosophical Essays, R. Jeffrey (ed.), Cambridge, Cambridge University Press, 1999, 229–49.

Hoffman-Grüneberg, Frank (1988), Radikal-empirische Wahrheitstheorie. Eine Studie über Otto Neurath, den Wiener Kreis und das Wahrheitsproblem, Vienna, Verlag Hölder-Pichler-Tempsky.

Kokoszyńska, Maria (1936a), 'Syntax, Semantik und Wissenschaftslogik', Actes du Congrès International de philosophie scientifique, vol. 3, Paris, Hermann, 9–14.

Kokoszyńska, Maria (1936b), 'Über den absoluten Wahrheitsbegriff und einige andere semantische Begriffe' Erkenntnis 6, 143–165.

Kokoszyńska, Maria (1937–1938), 'Bemerkungen über die Einheitswissenschaft', Erkenntnis 7, 325–335.

Mancosu, Paolo (2005), 'Harvard 1940–1941, Tarski, Carnap and Quine on a finitistic language of mathematics for science', History and Philosophy of Logic 26, 327–357.

Mancosu, Paolo (2006), "Tarski on models and logical consequence", in J. Ferreiros, J. Gray (eds.), The Architecture of Modern Mathematics, Oxford, Oxford University Press, 209–237.

Mancosu, Paolo (2008), 'Neurath, Tarski and Kokoszyńska on the semantic conception of truth', in D. Patterson (ed.), Essays on Tarski. Oxford, Oxford University Press, 192–224.

Mostowski, Andrzej (1967), 'Alfred Tarski', in Encyclopedia of Philosophy, Paul Edwards (ed.), vol. 8, New York, McMillan.

Neurath, Otto (1936), 'Erster Internationaler Kongress für Einheit der Wissenschaft in Paris 1935', Erkenntnis, 377–430.

Poznański, Edward and Alexander, Wundheiler (1934), 'Pojecie prawdy na terenie fizyki' (The concept of truth in physics), in Fragmenty Filozoficzne, Warsaw, 97–143. A partial German translation by Rose Rand in Carnap's Nachlass is found under call number RC 081-37-01.

Rodriguez-Consuegra, F. (2007), 'Two unpublished contributions by Alfred Tarski', History and Philosophy of Logic 28, 257–264,

Rojszczak, Artur (2002), 'Philosophical background and philosophical content of the semantic definition of truth', Erkenntnis 56, 29–62

Suppes, Patrick (1988), 'Philosophical implications of Tarski's work' The Journal of Symbolic Logic 53, 80–91.

Tarski, Alfred (1932), 'Der Wahrheitsbegriff in der Sprachen der deduktiven Disziplinen', Anzeiger der Akademie der Wissenschaften in Wien 69, 23–25. Reprinted in Tarski 1986, vol. I.

Tarski. Alfred (1933), 'Pojęcie prawdy w językach nauk dedukcyjnych', (The concept of truth in the languages of deductives sciences), Prace Towarzystwa Naukowego Warszawskiego, wydzial III, no. 34.

Tarski, Alfred (1935a), 'Der Wahrheitsbegriff in den formalisierten Sprachen', Studia Philosophica (Lemberg) 1, 261–405. Reprinted in Tarski 1986, vol. II. Eng. trans. in Tarski 1956, 152–278

Tarski, Alfred (1944), "A philosophical letter of Alfred Tarski' with a prefatory note by Morton White', Journal of Philosophy 84(1987), 28–32.

Tarski, Alfred (1956), Logic, Semantics, Metamathematics, Oxford, Oxford University Press. Second edition, 1983.

Tarski, Alfred (1986), Collected Papers, S. Givant and R. McKenzie (eds.), vols. I–IV, Basel, Birkhäuser.

Tugendhat, Ernst (1960), 'Tarskis semantische Definition der Wahrheit und ihre Stellung innerhalb der Geschichte des Wahrheitsproblems im logischen Positivismus', Philosophische Rundschau 8, 131–159; reprinted in G. Skirbekk, Wahrheitstheorien, Frankfurt, Suhrkamp, 1977, 189–223.

Uebel, Thomas (1992), Overcoming Logical Positivism from within. The emergence of Neurath's Naturalism from the Vienna Circle's Protocolol Debate. Amsterdam, Rodopi.

Woleński, Jan (1993), 'Tarski as a philosopher', in F. Coniglione, R. Polis and J. Woleński (eds.), Polish Scientific Philosophy, Amsterdam, Rodopi, 319–338.

Woleński, Jan (1995), 'On Tarski's background', in J. Hintikka (ed.), *From Dedekind to Gödel*, Dordrecht, Kluwer, 331–341.

Woleński, Jan and Eberhard Köhler (1998), *Alfred Tarski and the Vienna Circle*, Dordrecht, Kluwer.

Chapter 10
Tarski on Definition, Meaning and Truth

Douglas Patterson

The conception of truth was a central concern of Polish philosophy. Though there was some disagreement about it,[1] what was known as the "classical Aristotelian" conception was the dominant view. Tarski's presentation of his conception of the conditions under which this conception could rigorously be expressed—his Convention T—and his development of a method for expressing truth so conceived in a range of cases was the culmination this venerable tradition. Tarski's achievement is not without its detractors today; in particular Putnam's remark that "as a philosophical account of truth, Tarski's theory fails as badly as it is possible for an account to fail" (Putnam 1994, 333) is probably one of the more familiar evaluations of Tarski's contribution on record. Here, I will set out Tarski's conception of truth in its relation to Polish philosophy and other developments of the time; I will then respond to Putnam's criticism in the hope of vindicating Tarski and the tradition to which he belonged.

10.1 Tarski's Tripartite Conception of Meaning

Tarski's conception of meaning has three strands, which I will call the conceptions of *semantic meaning*, *formal meaning*, and *intuitive meaning*. The intuitive meaning of a term is, as Tarski has it, the concept expressed by it, while its formal meaning is determined by its role in a deductive theory; the semantic meaning is, in turn, the extension determined by its formal meaning if all goes well. In order to understand Tarski's definition of truth, then, we need to keep track of all three notions of meaning, and we need also to understand Tarski's account of the intuitive meaning of "true", the bit of philosophy that animates the formal work: what he calls the "classical Aristotelian" conception of truth. Here I'll discuss the three aspects of meaning, the first two only as much as necessary for present purposes; the reader wanting a fuller treatment may consult Patterson (2008a).

D. Patterson (✉)
Kansas State University, Manhattan, KS, USA
e-mail: pattersd@ksu.edu

[1] See Kijania-Pacek, this volume, on some dissenters.

S. Lapointe et al. (eds.), *The Golden Age of Polish Philosophy*, Logic, Epistemology, and the Unity of Science 16, DOI 10.1007/978-90-481-2401-5_11,

Intuitive Meaning. In a way that ought to strike interpreters as noteworthy more often than it has, Tarski often insists that the very "notions" he labors to define are "clear", "unambiguous" or "intuitive". For instance, Tarski is happy to say at the outset of "The Concept of Truth in Formalized Languages" that

> A thorough analysis of the meaning current in everyday life of the term "true" is not intended here. Every reader possesses in greater or less degree an intuitive knowledge of the concept of truth and he can find detailed discussions on it in works on the theory of knowledge.
>
> (1983, 153)

Likewise, in various other discussions, often before embarking on an involved project of formal definition, Tarski insists that the "intuitive meaning" or "concept expressed by" the definiendum is in some way perfectly clear. Before discussing semantic definition, which Tarski understands in terms of satisfaction, he says of the latter:

> ... we can try to define the sense of the following phrase: "*A finite sequence of objects satisfies a given sentential function.*" The successful accomplishment of this task raises difficulties which are greater than would appear at first sight. However, in whatever form and to whatever degree we do succeed in solving this problem, the intuitive meaning of the above phrase seems clear and unambiguous.
>
> (1983, 117)

Likewise, "The Semantic Conception of Truth" states its aim with respect to the "notion" of truth in this familiar passage:

> Our discussion will be centered around the notion of *truth*. The main problem is that of giving a *satisfactory definition* of this notion, i.e. a definition which is *materially adequate* and *formally correct*... The desired definition does not aim to specify the meaning of a familiar word used to denote a novel notion; on the contrary, it aims to catch hold of the actual meaning of an old notion. We must then characterize this notion precisely enough to enable anyone to determine whether the definition actually fulfills its task.
>
> (1944, 341)

This theme of rendering intuitions precise, related to use of the phrase "intuitive meaning" also receives extended discussion in the case of semantic definability:

> The problem set in this article belongs in principle to the type of problems which frequently occur in the course of mathematical investigations. Our interest is directed towards a term of which we can give an account that is more or less precise in its intuitive content, but the significance of which has not at present been rigorously established, at least in mathematics. We then seek to construct a definition of this term which, while satisfying the requirements of methodological rigour, will also render adequately and precisely the actual meaning of the term. It was just such problems that the geometers solved when they established the meaning of the terms "movement", "line", "surface", or "dimension" for the first time. Here I present an analogous problem concerning the term "definable set of real numbers".
>
> Strictly speaking this analogy should not be carried too far. In geometry it was a question of making precise the spatial intuitions acquired empirically in everyday life, intuitions which are vague and confused by their very nature. Here we have to deal with intuitions more clear and conscious, those of a logical nature relating to another domain of science, metamathematics. To the geometers the necessity presented itself of choosing one of several incompatible meanings, but here arbitrariness in establishing the content of the term in question is reduced almost to zero.

> I shall begin by presenting to the reader the content of this term, especially as it is now understood in metamathematics. The remarks I am about to make are not at all necessary for the considerations that will follow—any more than empirical knowledge of lines and surfaces is necessary for a mathematical theory of geometry. These remarks will allow us to grasp more easily the constructions explained in the following section and, above all, to judge whether or not they convey the actual meaning of the term.
>
> (1983, 112)

Tarski thus holds that "concepts grasped in everyday life" or "intuitive" "notions" or "meanings" are clearly enough understood at the outset of his inquiries that an ordinary grasp on them suffices for an ability to evaluate how well his definitional projects succeed.

Hence "intuitions", "intuitive meaning", ordinary "concepts" or "notions" are taken by Tarski to be largely perspicuous, especially in the cases in which we're interested here, and guide the setting out of formal definitions.[2] Indeed, Tarski retains this talk even in cases where formally correct definition is impossible. Although "Every reader possesses in greater or less degree intuitive knowledge of the concept of truth", nevertheless:

> In §1 colloquial language is the object of our investigations. The results are entirely negative. With respect to this language not only does the definition of truth seem to be impossible, but even the consistent use of this concept in conformity with the laws of logic.
>
> (1983, 153)

It follows that according to Tarski there is a concept of truth that cannot be used "in conformity with the laws of logic" in its application to colloquial language (because of the expressibility therein of the antinomy of the Liar; see Patterson 2006). Thus there is more to the concept of truth than is captured in any particular formal definition for a language, since the concept remains even in cases where the definition is impossible. It is likewise clear from these passages that concepts, unlike the meanings of terms that express them, are not relative to a language or a deductive theory; the concept of truth is independent enough of the meaning of "true" in a particular language to allow us to evaluate attempts at its definition, even when the evaluation is ultimately that definition is impossible. Historically, this line of thought in Tarski's remarks on meaning seems to be the remnant of what he endorses in one place as Leśniewski's "intuitionistic formalism", which seems to have amounted to the view that one ought, despite formalization, regard the terms used in a formalism as having meanings one grasps *independently* of the formalization (1983, 62).[3]

Formal Meaning. Let us now turn to Tarski's conception of *formal meaning*, a conception which allies Tarski directly with Hilbert and logical positivists such as Carnap. The primary text here is "Some Methodological Investigations on the

---

[2] It is in this respect that Tarski may still be influenced by Brentanian doctrines about the intuitive evidence of meaning, as Woleński and Simons (1993) assert. I was too quick to dismiss this suggestion in 2006.

[3] Unfortunately there is nothing illuminating in the passage from Leśniewski Tarski cites. See Hodges (2008) for one interpretation. In a note added to the page for the 1956 edition Tarski says that he ceased to endorse the Leśniewskian conception.

Definability of Concepts" and its discussion of formal definition in terms of the derivability of an explicit definition of the *definiendum* from a theory:

> Every sentence of the form:
>
> $$(x) : x = a. \Leftrightarrow .\phi(x; b', b'', \ldots)$$
>
> where "$\phi(x; b', b'', \ldots)$" stands for any sentential function which contains "x" as the only real variable, and in which no extra-logical constants other than "b", 'b"', ... of the set B occur, will be called a *possible definition* or simply a *definition of the term "a"by means of the set B*. We shall say that the term "*a*"*is definable by means of the terms of the set B on the basis of the set X of sentences*, if "*a*" and all terms of B occur in the sentences of the set X and if at the same time at least one possible definition of the term "*a*" by means of the terms of B is derivable from the sentences of X. (1983, 299)

Note that formal definability is always relative to a theory, the set of sentences X. Case studies here include, for example, the number of primitives required to express geometry (1983, 306).[4] We can see the importance of this conception of a term's meaning as determined by sentences involving it that are held true in a theory in Tarski's discussion of the fact that eliminability is always relative to a theory:

> there is no sense in discussing whether a term can be defined by means of other terms before the meaning of those terms has been established, and on the basis of a deductive theory we can establish the meaning of a term which has not previously been defined only by describing the sentences in which the term occurs and which we accept as true.
>
> (1983, 299)

An explicit definition codifies the meaning of a term as established by the sentences containing it which may be derived within a theory: sentences that suffice to settle the meaning of these other terms also settle the meaning of the defined term when supplemented only by the definition. The point in formal definition is thus to establish that a theory involving a certain term is equivalent to some theory involving strictly fewer primitive terms supplemented only by the definition (1983, 306).

Now this way of thinking of definitions doesn't sit particularly well with a common view of definitions as somehow without content: a formally correct definition, though conservative over the theory to which it is added, and though it eliminates the defined term relative to the theory, makes all the difference between a sub-theory and a full theory and thus has whatever content the first lacks and the second has.[5] The extendability of one theory to another through formal definition doesn't show *all by itself* that the second has some particular status assumed to hold of the first.

---

[4] Here Tarski is most influenced by Veblen and the American Postulate Theorists such as Langford and Huntington, as well as by Hilbert. See Scanlan 2003 for the connection to the former which included, as Scanlan discusses, Tarski's intensive study of Langford's work in a seminar from 1927 to 1929. For the connection with Hilbert, see Sinaceur 2001 and 2008.

[5] See Hodges (2008) for an excellent discussion of the two conceptions of definition and their history as it relates to the interpretation of Tarski.

Like Tarski's conception of intuitive meaning, Tarski's conception of definition itself seems to have been inherited at least in some respects from Leśniewski (see Hodges, 2008): Leśniewski held that definitions should be "creative" in the sense that he conceived of them not as rules for abbreviating certain strings of symbols in a given deductive theory, but rather as the key act in producing a *new* deductive theory. Tarski's treatment of formal definition retains some Leśniewskian elements; a definition isn't a convention for rewriting other sentences in shorthand, but is rather a sentence in itself like any other; its interest is that it plays a deductive role within a theory that makes clear that a certain intuitive meaning is in fact expressed by the *definiendum* in its deductive role within the theory. That a sentence plays the role of a definition relative to a theory is a matter of eliminability and conservativeness relative to that theory, as the standard theory of explicit definition developed following Tarski (Belnap 1993) holds today, but Leśniewskian "creativity" remains in the emphasis on expressing intuitive notions through making explicit through a single sentence the deductive role of an expression in a theory.

Tarski's conception of what I am calling "formal meaning" is also familiar from positivist views about the implicit definition of theoretical terms. Consider, for instance, Carnap, who writes, in 1934:

> Up to now, in constructing a language, the procedure has usually been, first to assign a meaning to the fundamental mathematico-logical symbols, and then to consider what sentences and inferences are seen to be logically correct in accordance with this meaning. Since the assignment of the meaning is expressed in words, and is, in consequence, inexact, no conclusion arrived at in this way can very well be otherwise than inexact and ambiguous. The connection will only become clear when approached from the opposite direction: let any postulates and any rules of inference be chosen arbitrarily; then this choice, whatever it may be, will determine what meaning is to be assigned to the fundamental logical symbols. (2002, xv)[6]

This conception of meaning is thus familiar from thinkers whose influence on Tarski is well-documented and can be found expressed in passages from Tarski like the one presented above.[7] The basic idea is that within a formal theory, some axioms are chosen and taken to be such that they "may be asserted" or "are considered as true" and the terms involved are taken to have meanings that are settled by their deductive role within the theory.[8]

It is very easy to confuse this idea with the idea that the terms have meanings that *make the axioms true*, so we should pause here to emphasize the difference. Tarski's conception of formal meaning is one on which a term gets its meaning from its role in those sentences that are *treated as true* in a theory. Tarski does not,

---

[6] I think this Carnapian attitude of "tolerance" is part of what is behind the remarks on rival conceptions of truth in the second part of Tarski 1944.

[7] There are relations here to Hilbert as well. For Hilbert, see e.g. Tarski 1941, 120, 140, also Sinaceur 2001 and 2008 , and for Carnap see the many favorable references to the *Abriß der Logistik* in Tarski 1983.

[8] See Detlefsen 2004 and Coffa 1986 (both Detlefsen and Coffa trace the view back much further than Hilbert) as well as Friedman 1999 for related discussion.

on my reading, subscribe to the additional thesis, sometimes attributed to Hilbert (at least where *consistent* mathematical theories are concerned) that the terms of a theory are endowed by their role in the theory with meanings that guarantee the truth of the axioms of the theory. On Tarski's conception, it is perfectly possible for terms to have their formal meanings determined by their roles in a false theory. This point about the relation between formal meaning and truth will be relevant when we reject the commonly accepted claim that if Tarski's definition of truth were correct the expressions of the object language would have to have necessarily the meanings that they actually have.

This positivist, inferential conception of meaning ties meaning to a language only via the mediation of what is derivable from what there (including what is treated as derivable from no premises, that is, as axiomatic) and this, in turn, explains a fact often noted with some consternation by interpreters, that Tarski often speaks freely of languages as individuated by the theories consisting of their assertible sentences and as having properties, such as "inconsistency", that only sets of sentences can have (Tarski, 1983, 165; 1944, 349).[9] Inconsistency of a language comes to nothing more than the derivability of contradictions from sentences that determine meanings.

Semantic Meaning. Finally, we need to look at Tarski's remarks on the role of extension—satisfaction, denotation, and so on—in meaning; this constitutes the third strand in his tripartite conception. Semantic definition, the definition of extensions by meaningful terms, is the topic of "On Definable Sets of Real Numbers". The main sort of question about this sort of definition asks whether, given a language and an interpretation of its primitive vocabulary, various constructs out of this interpretation are in fact the extensions of expressions constructed out of the vocabulary via whatever formation rules are provided.

In a way that strongly anticipates the fundamental negative results on the definability of truth, Tarski shows how, given an association between certain primitive sentential functions of a system sufficient for the arithmetic of the real numbers and the sets that satisfy them, the set of all arithmetically definable sets of reals can be defined recursively in non-semantic terms. As he notes, however, it follows, on pain of a contradiction in the form of the Richard paradox that this set does not include all sets of reals (1983, 119). Tarski relates truth to the satisfaction of a sentence by all sequences of objects in a way that parallels exactly the definition of truth in "The Concept of Truth in Formalized Languages" (1983, 117), so that in its essentials the celebrated definition of truth of the long article is already present in the shorter treatment of semantic definability. Since the more famous article goes into significantly more detail on the definition of truth, we can expect that merely

[9] Of course, this alone doesn't justify including full type theory, systems sufficient for arithmetic, and so on, under the heading of "language", but as is often noted, during the period under study here Tarski often seemed happy to include these things under the general heading of "logic" (see Feferman (2008)), and this assimilation would have made it more natural to attach the relevant deductive systems to "languages". Furthermore, to the extent that Tarski was attracted to formalist doctrines about implicit definition, the assimilation would also seem natural to him.

introducing an expression that defines the set of truths cannot have been Tarski's only aim there. What is added, as we'll see, is an extended discussion of the attempt to capture the *intuitive meaning* of semantic concepts in a certain sort of formal definition.

There are a number important connections between the conception of meaning as settled by sentences held true discussed above and the conception of meaning as extension in Tarski's work. For our purposes here, however, it will do to note that for Tarski, a term's formal meaning will, under favorable circumstances, determine a semantic meaning for it, but that circumstances will not always be favorable, and that when they are not Tarski, in the seminal papers of the late 1920s and early 1930s, prefers to cling to formal meaning at the expense of semantic meaning. This can be seen in his suggestion that even when an explicit definition of truth cannot be given, we can develop the theory of truth axiomatically by taking the T-sentences as axioms; in so doing, as Tarski notes in the discussion of Theorem II of "The Concept of Truth in Formalized Languages", we give up on full determination of extension (since the resultant theory cannot be categorical) but we retain the basic deductive role of the truth-predicate, the one which is essential to its expressing the classical Aristotelian conception of truth. See Patterson 2006 and 2008 a for more on these topics; the point here that what is of most interest to Tarski in the 1930s is whether the intuitive notion of truth can rigorously be expressed in an expression taken up in a deductive theory; formal semantics as the theory of extension, though it flows out of Tarski's work on semantic definition, was in the seminal papers actually secondary to his concern with the intuitive meaning of "is true". We will now turn directly to a discussion of Tarski's remarks on this central topic.

10.2 The Aristotelian Conception of Truth in Tarski's Definitional Project

When Tarski sets out to define truth, we should, on the basis of the above considerations, expect him to have all three strands of the conception of meaning in mind: we should take him to be looking to express an *intuitive* conception of truth with a term endowed with a *formal* meaning by its role in a deductive theory, and we should expect him to be interested in the *semantic* extension thereby determined for the term. As for the "intuitive meaning" of "is true", Tarski describes his goal as being to find a "precise expression" (1944, 343) of the "intuitions which adhere to the *classical Aristotelian conception of truth*" (1944, 342). On this conception—which, note, Tarski associates not with his formal definitions, but with the concept which those definitions are intended to express—a good definition of truth is one according to which

> (1) *a true sentence is one which says that the state of affairs is so and so, and the state of affairs indeed is so and so.*
>
> From the point of view of formal correctness, clarity, and freedom from ambiguity of the expressions occurring in it, the above formulation obviously

leaves much to be desired. Nevertheless its intuitive meaning and general intention seem to be quite clear and intelligible. To make this intention more definite, and to give it a correct form, is precisely the task of a semantical definition.

As a starting-point, certain sentences of a special kind present themselves which could serve as partial definitions of the truth of a sentence or more correctly as explanations of various concrete turns of speech of the type "*x* is a true sentence". The general scheme of this kind of sentence can be depicted in the following way:

(2) *x is a true sentence if and only if p.*

In order to obtain concrete definitions we substitute in the place of the symbol "*p*" in this scheme any sentence, and in the place of "*x*" any individual name of this sentence. (1983, 156–157)

This is Tarski's philosophical analysis of the notion of truth. Formal definitions are beholden to it in that they'll be evaluated as successful to the extent that they can be seen as expressing this conception. Thus the intuition to be rendered precise is that a sentence "p" is true if and only if p, that is, that what are commonly known as the "T-sentences" settle the meaning of "is true" (1983, 165 187, also 1944, 348). This intuitive conception is a conception as to which sentences involving an expression "is T" are to be treated as theorems in a deductive theory if "is T" is to be regarded intuitively as expressing the notion of truth in the language of that theory. The semantic conception of truth is thus a conception of what conditions the formal meaning of a term must satisfy if the term is to express a certain intuitive meaning. The requirement that the T-sentences be implied by a definition (relative to the theory to which it is added) is thus a philosophical requirement based on the classical Aristotelian conception of truth, and is not a merely formal requirement that grows out of the mere intention properly to capture the *extension* of "is true" (see Patterson 2006a).

Woleński and Murawski (2008) discuss how this conception of truth had a long history in Polish philosophy, in the work of Twardowski and Kotarbiński, from whom Tarski took many of his basic ideas about truth, and in particular the idea that the Aristotelian formulation we have examined is the centerpiece of the proper treatment of truth. They also note that Czeżowski seems to have been the first author to focus on what we now know as the T-schema as a way of working out the correct conception of truth. As Hodges (2008) notes, the *name* "semantic conception of truth" comes from Kotarbiński's treatment of *semantic definitions* in his *Elementy*; for Kotarbiński such definitions are definitions in which the *definiendum* is *mentioned* but the *definiens* is *used*, and Kotarbiński distinguishes them from definitions that uniformly use, or uniformly mention, symbols. Hodges goes on to discuss in detail the way in which Kotarbiński's treatment of these topics played a crucial role in Tarski's discovery of a way of turning his ongoing work on quantifier elimination (see here Hodges (2008), as well as Scanlan 2003) into the method of defining truth that made him "a giant in the world of ideas" (Hodges, 2008).

The details of this procedure are now familiar enough that I'll simply give the briefest of summaries here.[10] The definition is to be a definition in a metalanguage *ML* for the sentences of an object language *L*. One first defines the satisfaction of an open sentence of *L* by a sequence of objects, calling an open sentence satisfied by a sequence just when the relevant members of the sequence stand in the relation expressed by the open sentence. Saying this in any particular case involves *using* an expression that translates the predicative component of the open sentence. In Tarski's own example, there is only one lexical predicate, "$\subseteq$", so the relevant clause is, notational niceties aside: "$x \subseteq y$" is satisfied by a sequence including x and y iff x is a subset of y. Sentential connectives are handled in the familiar way (e.g., an open sentence with "or" as its main connective is satisfied just in case either one or the other disjoined open sentences is satisfied, etc.) and quantification is handled by looking at preservation of satisfaction across variations in sequences at the relevant positions (e.g., "there is an x such that $x \subseteq y$" is satisfied by a sequence iff "$x \subseteq y$" is satisfied by at least one of the sequences that differ from the sequence in question only with respect to what they assign to "x") (Def 22, 1983, 193). Suitable use of higher-order logic or set theory turns these recursive conditions into an explicit definition for membership in a set. A true sentence is then defined as a sentence satisfied by every sequence (Def 23, 1983, 195). Since the definition of satisfaction makes liberal use of expressions of *ML* that translate the corresponding expressions of the *L*, the result is a definition that implies the T-sentences for *L* in *ML*. (1983, 195–196).

Before moving on, we need to set aside one very familiar worry about Tarski's procedure, namely that the criterion of adequacy that expresses his philosophical conception of truth, Convention T, *uses* the semantic notion of translation and that this is somehow incompatible with his goal of eliminating semantic terms (1983, 152–153). Convention T states a sufficient condition for the "material adequacy" of a definition of "is an element of the set of true sentences". As Tarski himself notes, Convention T, strictly speaking, belongs to a metalanguage *MML* for *ML* itself. As such, it is a claim in *MML* about the conditions under which a formal definition of this expression in *ML* will be a *good* one. But this goodness can, in the end, only be explicated *using* the notion of truth in *MML*: a good definition of truth for *L* in *ML* is one that introduces a predicate that has as its extension the *truths* of *L* (given the extra-linguistic facts, of course). The worry, then, is about that unexpunged use of "true" in the statement, in *MML*, of the conditions under which a definition in *ML* will be treated as acceptable, or is, more often, to the effect that the notion of truth has somehow illegitimately been smuggled into the definition of truth concealed in the notion of translation, since good translation would seem to require at least the preservation of truth conditions.

However, only a confusion about definition makes it look as though there is anything suspect here. The *evaluation* of a definition as *good* or not, as "adequate" or "accurate", or the evaluation of a certain predicate with respect to its having the "right" or "intended" extension, takes place *not* in the language in which the

[10] See Soames 1999 for one of many accessible treatments.

definition is stated, but in a metalanguage for it. Hence in Tarski's case, the criterion of material adequacy on a definition of truth belongs in *MML* rather than *ML*. For a simple example that leaves aside the complicated three-language structure of Tarski's adequacy criterion, consider the common definition of a bachelor as an unmarried man. When we define a bachelor as an unmarried man, we want to say that this definition is *correct*. But this can only mean that it is really the case that all and only unmarried men are in fact *bachelors*. We can't justify the claim that "bachelors are unmarried men" is a good definition of "bachelor" without ourselves saying "bachelor". There is nothing circular about this, and likewise, there will be nothing circular in saying, *of* a definition of truth, that it applies to all and only certain sentences that are *true* or that it expresses the concept of *truth*. Such a claim relates the intuitive meaning of "is true" to an expression in a formal language intended to have a formal meaning that is beholden to it. The evaluative claim is a claim about a definition made in a metalanguage; it is not somehow a circular element taken up in the definition itself (see Patterson 2007). Hence there could be no circularity in Tarski procedure even if he did simply say that a materially adequate definition of "is true" has to imply certain sentences involving it that are *true*, and therefore there is no circularity in his appeal to the notion of translation in the statement of Convention T, even if translation is a matter, at least in part, of preservation of *truth* conditions.

The question as to whether a definition of "is true" is "materially adequate" is the question whether the intended meaning of the expression—its "intuitive meaning" or the "concept" or "notion" that is supposed to be expressed by the expression of *ML*—is actually present in the set of sentences held true in a theory, but the question can't be asked without saying what the meaning is, and this in turn can't be done without using an expression of *MML* that is assumed to have the intended meaning. In Tarski's case this appeal to intuitive meaning comes in the assumption that when one sentence translates another it also states its truth condition, the assumption without which Convention T makes no sense. We can't say of Def. 23 that it gets it *right* unless, apprised of the definition of satisfaction we're willing to assent to the claim that sentences satisfied by all sequences are in fact *true*. But the simple equation of truth with "satisfaction by all sequences" is hardly going to effect this; its the link to the T-sentences that renders the definition intuitively satisfying (relative, that is to the Aristotelian conception of truth). If, with Tarski, we accept this conception, then we can recognize, of an expression defined so as to imply the T-sentences, that it expresses the relevant concept, since we are ourselves willing to accept that, where what's substituted for "p" translates *s*, the sentence *s* is true if and only if p. There is no circularity here; indeed, there would be no circularity in saying that it is a criterion of adequacy on a definition of truth that it imply those instances of "*s* is true in L if and only if p" that are in fact *true*.[11] This, by the way, is also what is missing from the otherwise parallel definition of truth in "On Definable Sets of Real Numbers": in "The Concept of Truth in Formalized

[11] Putnam betrays the common confusion that this would somehow be circular at 1994, 319–320.

Languages" the same formal definition is explicitly evaluated for its capturing the intuitive notion of truth via the T-sentences. What's added to a definition of truth already present but not treated as particularly interesting in the earlier article is an explicit argument to the effect that the definitions given capture not just *extension* but *intuitive meaning*.[12] It follows that material adequacy and extensional adequacy are not the same, despite common assumptions to the contrary. The remarks at 1983, 129 are perfectly clear that "material adequacy" is a matter of intuitive meaning as opposed to merely correct determination of extension, for instance.

Convention T is thus Tarski's expression of the "classical Aristotelian" conception of truth he inherited from his teachers in the Lvov-Warsaw school. It is a substantial, philosophical account to be expressed in *MML* of the conditions under which a definition of truth for *L* in *ML* is "materially adequate", that is, the conditions under which it "catches hold of the actual meaning of an old notion". This philosophical account, and its ultimately unreduced deployment of the notion of truth, is expressed in a language in which we evaluate particular definitions and is not itself wholly taken up in any one definition. This shows both, as we have seen, how the account doesn't render the definitions circular, and it also shows, as we will see next, how the definitions can't be expected to do all of the work the account itself does.

10.3 Standard Objections Debunked

Attention to the details of Tarski's conception of meaning corrects a number of stock criticisms of Tarski. To focus discussion, let us consider these familiar claims from Putnam:

> Since (2) [snow is white' is true-in-L iff snow is white] is a *theorem of logic* in meta-L (if we accept the definition, given by Tarski, of "true-in-L") since no axioms are needed in the proof of (2) except axioms of logic and axioms about spelling, (2) holds in all possible worlds. In particular, since no assumptions about the *use* of the expressions of *L* are used in the proof of (2), (2) holds true in worlds in which the sentence "Snow is white" does not mean that snow is white.
>
> ...all a logician wants of a truth definition is that it should capture the *extension* (denotation) of "true" as applied to L, not that it should capture the *sense*—the intuitive notion of truth (as restricted to L). But the concern of philosophy is precisely to discover what the intuitive notion of truth is. As a philosophical account of truth, Tarski's theory fails as badly as it is possible for an account to fail. A property that the sentence "Snow is white" would have (as long as snow is white) no matter how we might use or understand that sentence isn't even doubtfully or dubiously "close" to the property of truth. It just isn't truth at all.
>
> (1994, 333)

One idea here is that since Tarski makes "'snow is white' is true-in-L iff snow is white" a *definitional* truth, it is thereby according to his definition a *necessary* truth, whereas it is obviously a contingent fact that the sentence has that truth-condition,

[12] Some of the remarks in Patterson 2006a now strike me as confused on the difference.

since the truth condition of a sentence depends on the meanings and ultimately the uses of its expressions and expressions have their meanings contingently because these uses are themselves contingent. Tarski is therefore guilty of turning obviously contingent truths into necessary truths and his procedure is thereby deeply flawed.

In order to evaluate the criticism, we must appreciate first, though Tarski does not emphasize the fact in the writings of the period, that a formal definition is relative not only to a language and a theory, but to a set of contexts as well (Belnap 1993, 121), and that in Tarski's case the set of contexts is always the extensional contexts. Definability of a term relative to a language and theory in extensional contexts need not imply that it is definable in intensional contexts, should the language and theory include these.[13] The point is easy to miss since Tarski is always interested in theories formulated in extensional languages, but it nevertheless crucially affects the interpretation of his definitions. What Putnam's complaint ignores is that Tarski's definition eliminates the truth-predicate *in extensional contexts only* and is thus neutral as to necessity or contingency, since Tarski's background theories themselves have no resources to distinguish necessary from contingent truth.

It is thus unwarranted to infer from the fact that the T-sentence follows from axioms for formal syntax *plus* Tarski's definition that it, in the language in which it is given, expresses something that "holds in all possible worlds". It is even more unwarranted to claim, as Putnam and others do, that according to Tarski Putnam's (2) "is a theorem of logic"—granting, temporarily, the use of the term "logic" to cover Tarski's use of higher-order type theory plus a theory of formal syntax. It isn't, and Tarski never said it was: it is a theorem of *"logic" so construed plus the truth-definition.* The truth definition itself is not a truth of "logic", that is, it is not a truth of higher-order type theory plus formal syntax, any more than "bachelors are unmarried men" is a truth of logic. It is clear from our discussion above that though Tarski holds that definitions must be eliminable and conservative, he does *not* hold that they're somehow without content so that anything that follows from a theory with them added is somehow without further assumption of the same status as the theorems of the unsupplemented theory. On the contrary: showing a term to be eliminable via a definition relative to a sub-theory of a theory is showing that the sub-theory can be conservatively enriched to the whole theory by the definition alone; it *isn't* showing that there is no difference in status between the sub-theory and the theory. In particular, even if Tarski did think that the logical and mathematical truths of the background theory of *ML* were necessary truths—though he never says anything of the sort, in accord with the general Polish aversion to intensional contexts, and though, as emphasized above, nothing in the theories themselves forces this understanding of them—it doesn't follow that he thought his definitions were necessarily true, and hence it doesn't follow that he thought

[13] Notice that in Putnam this is buried under the bizarre assumption that "logicians" are only interested in extensions.

the T-sentences were necessarily true. Since T-sentences are so clearly *not* necessary truths, we shouldn't take Tarski to be guilty of the error of proceeding on the assumption that they are without compelling reason.

Often interpreters try to save Tarski from Putnam's criticism by arguing that on his view languages are individuated by the meanings of their terms. Given this, defining truth in a given language in a way tied essentially to the (actual) meaning of a certain list of terms in that language looks acceptable, even given the assumption—the one I have argued here is mistaken—that definitions express necessary truths. The view, however, can't make sense of the fact that Tarski clearly, and quite reasonably, allows that "as far as natural language is concerned... this language is not something finished, closed, or bounded by clear limits. It is not laid down what words can be added to this language and thus in a certain sense already belong to it potentially" (1983, 164). If Tarski allows that terms can be *added* to a language, then he cannot be thinking that it is a necessary truth about a language that it have exactly the lexicon it actually has. Of course Tarski holds that colloquial language cannot be treated rigorously for reasons such as unclarity as to what its terms are; the point is that we should not saddle him with the view that languages necessarily have exactly a certain set of terms with exactly certain meanings, as the standard defense does. Rather, I've suggested, we should recognize that definitions in extensional theories need only be true, not necessarily or logically true. (If it grates to hear definitions called "true", then just read: definitions need only allow the derivation of truths as opposed to necessary or logical truths.)

Now it might appear that this doesn't square well with the conception of formal meaning discussed above, for it might be thought that if a term has whatever meaning is required to make the axioms and theorems of some theory *true*, then there is no such thing as contingent truth: no such thing as a term preserving its meaning while a sentence the truth of which is supposed to determine its meaning goes from true to false. But this would be a confusion based on a misunderstanding of the conception of formal meaning: as was noted above, what determines meaning on this conception is not the *truth* of certain sentences, but their being *held true*, that is, their being treated as axioms and theorems. Carnap, for instance, doesn't say that the *truth* of "postulates" settles the meaning of terms; it is their being "chosen" to function as postulates that does so. A term's formal meaning is a matter of derivability relations in a formal theory, and Tarski's view is that we can't change *those* without changing the meaning of the term. This isn't unreasonable, amounting as it does merely to the view that a change in inferential role is a change in meaning.[14] Definitions could hardly be expected to work any other way.

A sentence's semantic meaning, that is, its truth condition, likewise needs to be settled by its formal meaning, since formal meaning is supposed to determine semantic meaning as discussed in Section 10.2. But a sentence's *truth value* is not likewise supposed on any view of Tarski's to be settled by our holding it

---

[14] Note that this unobjectionable, weak view leaves it open whether meaning determines inferential role or *vice versa*.

true.[15] Hence Tarski could allow that a theory, stated in *ML*, that has the T-sentences as theorems is such that the meaning of "is true in *L*" is settled within the theory *by that theory's having the axioms and deductive structure it has*. But this doesn't commit him to saying that the theory could not go from true to false (if *L* were to change) or that it could not have been false (had *L* been different), or for that matter that it could not *be* false (if, in fact, it has as theorems some instances of "*s* is true in *L* if and only if p" where what is substituted for "p" *doesn't* translate *s*).[16] These are claims, made in *MML*, about the *truth values* sentences of the theory expressed in *ML* might have or have had, and allowing that they are contingent is perfectly compatible with the idea that expressions of the theory expressed *in ML* mean what they mean (have the formal meanings they do) because of their deductive role within the theory. Now, of course, if we were to recognize that the meanings of the sentences of *L* had changed (or had them wrong in the first place), so that the *ML* truth-definition that implied them had gone false (or had always been so), we would have to change the definition in *ML*, and *that*, given Tarski's conception of formal meaning, would be a change in the formal meaning of "is true in *L*" in *ML*, since *different* sentences of the form "*s* is true in *L* iff p" would become the ones implied by the truth-definition. So Tarski is committed to the claim that we can't change the *definiens* in the definition of truth without changing the meaning of the *definiendum*. This, however, is not unreasonable: if we substitute in a definition a *definiens* with a new meaning, then the *definiendum* may of course be given a different meaning by the definition.

Turning next to the second paragraph of our quotation from Putnam, does Tarski's "philosophical" account of truth "fail as badly as it is possible to fail"? It is here that another widespread confusion, one which rests on failing to distinguish Tarski's remarks on intuitive meaning from those on formal meaning, and more generally rests on failing to note the difference between what is in *ML* and what is in *MML*, is at work. Tarski's formal definitions of truth predicates are not, and were not conceived by him to be, his philosophical account of truth. His philosophical account of truth is the "Aristotelian" conception, and the central plank of this account is the claim that a definition of truth should somehow sum up what is stated in the T-sentences. Whatever its merits on its own terms, this is a "philosophical account", and it is clearly *not* exhaustively expressed or intended exhaustively to be expressed in any *particular* formal definition of truth. Hence, of course, focus on the formal definitions alone will find them wanting as accounts of truth; the mistake, though, is to think that the philosophical view is exhaustively expressed in a definition, as opposed to the account of the conditions under which a definition is good. The philosophy takes place in *MML* as, ultimately, the account of

[15] I suppose we could worry about sentences that say of themselves that they're accepted by us, but these won't appear in formalized languages, and won't be true in natural languages because nothing is. The relevant issues are discussed in Patterson 2006.

[16] Though the last requires us not to take the idea of axioms as things that are "held true" too seriously, or the view is Moore-paradoxical; just read "held true" as "treated as derivable within the theory".

the conditions under which a definition in *ML*, relative to a set of contexts in *L* and a theory expressed in *L*, is a good one. Davidson (1990, 282–295) seems to me to come close to getting this right, though he underplays the importance of the Aristotelian conception and hence understates the extent to which Tarski did in fact tell us more about truth than his definitions do.

It follows also that we can ignore the very familiar complaint that Tarski's definitions don't tell us what they have in common as definitions of truth for various languages, and, relatedly, that they don't themselves explain how they are to be adapted to extensions of a language or to new languages: they aren't supposed to. (Making this charge is one respect in which Davidson 1990 doesn't get matters completely right.) Again, the Aristotelian conception in *MML* tells us under what conditions a definition is good—when it implies the T-sentences—and the interaction of this with facts about the syntax and semantics of *L* informs the construction of a definition for particular languages *L* in *ML*, whether they be entirely new, or related structurally or historically to other languages. It is perfectly clear here what is common to adequate definitions of truth: they all adhere to the classical Aristotelian conception of truth by implying (relative to their background theories) the T-sentences for the object languages they concern.

10.4 Conclusion

We thus see how careful attention to Tarski's tripartite conception of meaning allows us to be clear in ways that show that some very widely accepted criticisms are unjustified. Tarski's aim was to express the intuitive notion of truth, which he took to be characterized by the classical Aristotelian conception, in formal theories, thus achieving not only the aim of elimination of semantic terms in favor of those taken to be less problematic, and not only the determination of their extensions, but at least as importantly, the clear, formal expression of a certain philosophical conception of truth. The formal definitions answer to the Aristotelian conception of truth via Convention T. With these strands in the notion of meaning sorted out, we can see that there is no threat of circularity in Tarski's procedure, and that his definitions do not mistakenly take semantic truths for logical or necessary truths. We can also see that "Tarski's theory of truth" doesn't fail as a "philosophical account" in a way that his particular definitions taken as such would, because in being clear about the distinction between his account of the intuitive notion of truth and his formal definitions for particular languages we can see that "Tarski's theory of truth" isn't expressed in *any* particular formal definition, but is expressed, rather, in the adequacy condition for such definitions, Convention T.

Bibliography

Belnap, Nuel (1993), 'On Rigorous Definitions', Philosophical Studies 72, 115–146.

Coffa, Alberto (1986), 'From Geometry to Tolerance, Sources of Conventionalism in Nineteenth-Century Geometry', in Colodny, Robert (ed.), From Quarks to Quasars, Pittsburgh, University of Pittsburgh Press, 3–70.

Davidson, Donald (1990), 'The Structure and Content of Truth', The Journal of Philosophy 87, 279–328.
Detlefsen, Michael (2004), 'Formalism', in Shapiro, Stewart (ed.), Oxford Handbook of Logic and the Philosophy of Mathematics, Oxford, Oxford University Press, 236–317.
Feferman, Solomon 2008 'Tarski's Conceptual Analysis of Semantical Notions', in Patterson.
Friedman, Michael (1999), Reconsidering Logical Positivism, Cambridge, Cambridge University Press.
Hodges, Wilfrid 2008, 'Tarski's Theory of Definition', in Patterson 2008.
Patterson, Douglas (2005), 'Learnability and Compositionality', Mind and Language 20, 326–352.
Patterson, Douglas (2006a), 'Tarski on the Necessity Reading of Convention T', Synthese 151, 1–24.
Patterson, Douglas (2007), 'On the Determination Argument Against Deflationism', Pacific Philosophical Quarterly 88, 2.
Patterson, Douglas (ed.) (2008), New Essays on Tarski and Philosophy, New York, Oxford University Press.
Patterson, Douglas (2008a), 'Tarski's Conception of Meaning', in Patterson 2008.
Putnam, Hilary (1994), 'A Comparison of Something with Something Else', in Conant, James (ed.), Words and Life, Cambridge, MA, Harvard University Press.
Scanlan, Michael (2003), 'American Postulate Theorists and Alfred Tarski', History and Philosophy of Logic 24, 307–325.
Sinaceur, Hourya (2001), Alfred Tarski, Semantic Shift, Heuristic Shift in Metamathematics, Synthese 126, 49–65.
Sinaceur, Hourya (2008), 'Tarski's Practice and Philosophy, Between Formalism and Pragmatism', in Lindström, Palmgren, Segerberg and Stoltenberg-Hansen (eds.), Logicism, Intuitionism, and Formalism, What Has Become of Them? Dordrecht, Springer.
Soames, Scott (1999), Understanding Truth, New York, Oxford University Press.
Tarski, Alfred (1944), 'The Semantic Conception of Truth and the Foundations of Semantics', Philosophy and Phenomenological Research 4, 341–375.
Tarski, Alfred (1983), Logic, Semantics, Metamathematics, 2nd edition, J. Corcoran (ed.), New York, Oxford University Press.
Woleński, Jan and Murawski, Roman (2008), 'Tarski and his Polish Predecessors on Truth', in Patterson (2008). Woleński, Jan and Simons, Peter (1989), 'De Veritate, Austro-Polish Contributions to the Theory of Truth from Brentano to Tarski', in Szaniawski K.,1989, 391–442.

Part III
Polish Philosophy of Mind

Chapter 11
A Note on Henryk Mehlberg's Contribution to the Debate on the Mind-Body Problem

Urszula M. Żegleń

One classic question in the philosophy of mind is: What is the relation between mental and physical phenomena? Contemporary attempts at a solution to this problem typically reject substance dualism and tend towards monism, though this does not always mean reductive materialism. The problem of explaining how we are to account for mental states and their relations to physical phenomena (if there are such relations) or eliminate them altogether plays a central role and the options are many. We owe one the first comprehensive discussion of reductionism to the logical positivists of the Vienna Circle. But the philosophers of the Lvov-Warsaw School made significant contributions to this debate as well. Their own views, sadly enough, were not influential. Nonetheless, their analyses pertain to philosophical discussions that were lively and important at the time.

Henryk Mehlberg, who is otherwise known for his contribution to philosophy of science[1], was one of the most important contributors to the debate on "psychophysicalism", that is, the mind-body problem.[2] Mehlberg, under the influence of Charles D. Broad and Rudolf Carnap, considered the mind-body problem under the heading of "psychophysical parallelism". This, if we were to follow traditional terminology, would suggest that Mehlberg was interested in a form of dualism – the one most notoriously attributed to Leibniz – according to which mental events and physical events are of a different kind, so that while mental events may entertain causal relations with other mental events and physical events may cause other physical events, mental events cannot cause physical events and vice versa. However, Mehlberg's analyses of the traditional tenets of psychophysical parallelism suggest otherwise. He understood the question of parallelism to consist not in explaining how two distinct substances ontologically independent one of the other, the Mental and the Physical, happen to be somehow coordinated. Rather, he understood the problem

U.M. Żegleń (✉)
Nicolaus Copernicus University, Toruń, Poland
e-mail: zeglen@uni.torun.pl

[1] Mehlkberg taught in Chicago and then Toronto after he emigrated in 1945. He is the author of, among other titles, *Time, Causality and Quantum Theory*.

[2] Mehlberg, however, distinguished between the scientific problem of psychophysicalism and the traditional philosophical mind-body problem.

S. Lapointe et al. (eds.), *The Golden Age of Polish Philosophy*, Logic, Epistemology, and the Unity of Science 16, DOI 10.1007/978-90-481-2401-5_12,

to consist in explaining how what happens in the brain (which Mehlberg often designates as the "substratum") somehow "determines" – in a sense which need not be made precise at this point – what happens in the mind. For Mehlberg, the brain is substratum in the sense that it realises physical events which are simultaneous to mental experience and by which this experience is determined: what we find in Mehlberg's discussion is an account of the mind-body problem which is best portrayed as epiphenomenalism.

According to Mehlberg, psychophysical parallelism is characterized by three distinctive theses:

(T1) Correspondence (or Simultaneity): The sequence of a person's consecutive states of consciousness and the sequence of that person's consecutive brain states run "parallel" in time and stand in a relation such that brain states "determine" mental states (Mehlberg, *Parallelism*, 40), I'll come back to the inverted commas in what follows.

(T2) Causal independence of the mental: No mental phenomena cause or are the effects of, any physical phenomena.

(T3) Psychophysical identity: Mental phenomena that appear in introspection are in fact identical with nervous processes (ibid., 41). In other words, every mental phenomenon (experience) is numerically identical with its substratum, i.e., with a "correlative" physical phenomenon (ibid., 54).

The nature of Mehlberg's preferred variety of epiphenomenalism comes out of his discussion of these three (more or less consistent) theses, and of his discussion of the first one in particular. In what follows, and for the sake of brevity, I will therefore focus on (T1). Mehlberg's analyses led him to introduce a series of distinctions that are intended to explain what it means for a mental state to "correspond" to a brain state, and in particular to be "determined" by it. What is foremost interesting about his analyses is that they thrive – implicitly – on the assumption that in order to account for the correspondence between brain states and mental states one needs to account for the fact that that same mental state can be realized in a multiplicity of fashions. Although the conceptual framework and the terminology he uses are themselves at times somewhat bewildering, the theory he present was both original and historically groundbreaking. It was, at any rate, a clear anticipation of the well-known functionalist intuition.

How should we interpret (T1)? Mehlberg first considers the following construal:

(T1a) Every mental state (experience) is determined by a simultaneous state within the nervous system (Mehlberg, *Paralelizm*, 353), i.e., by its physiological "substratum" (Mehlberg 1980b, 270).

The first question to arise is whether (T1a) is saying something about types of states (believing, desiring, etc.) or tokens of such types (say, someone's being, at time *t*, in the state of believing that *p*). As Mehlberg sees it, if (T1a) were about types of states, then it would seem impossible to determine what kinds of experience are to be counted as believings, desirings, etc, otherwise than *a priori* and (T1a), then, would appear to be trivial (Mehlberg, *Paralelizm*, 325). As we shall see, Mehlberg opts for an interpretation on which the correspondence holds between tokens. Because it thrives on this ambiguity Mehlberg is, however, altogether dissatisfied with (T1a) and likewise rejects the following somewhat modified conditional version of (T1a)

(T1a') If a brain state determines simultaneous experiences, then intrinsically similar brain states are accompanied by intrinsically similar experiences.

But Mehlberg does not want to discard entirely the intuition on which (T1a') lies, and the ensuing analyses aim at clarifying the notion of "intrinsic similarity" it introduces. He begins by pointing that "intrinsic" in "intrinsic similarity" should neither be taken to mean "essentially" nor "absolutely", nor be defined in terms of objects sharing *all* properties (as in Leibnizian identity). Instead, Mehlberg maintains that objects that are intrinsically similar in the relevant sense need in fact share only *some* (and not all) determinate kinds of properties. What kinds of properties do intrinsically similar object share? Various examples lead Mehlberg to claim that, typically, the properties objects have in common when they are intrinsically similar have the particularity of being gradational in the sense that they change continuously. So, for instance, two individual auditory sensations are intrinsically similar on account of their sharing volume, timbre, pitch or duration (Mehlberg 1980a, 42) which are all (and also the only) gradational properties of auditory sensations. As Mehlberg sees it, both physical properties such as the pitch of a sound or the hue of a certain color – here Mehlberg clearly follows a Brentanian understanding of what counts as "physical" – as well as the Brentanian idea, also found in Carl Stumpf and the early Husserl that at least some type of perceptions are gradational and that relations of dependence between features of objects can be defined on the basis of the way in which the variation of one aspect affects the variation of another. For Mehlberg, continuity also applies to putative "internal" perceptions of mental properties such as the strength of conviction of someone's belief, or the intensity of the pleasure or pain associated with a particular emotion. The perception of such properties is also gradational.

Of course, the idea that sensory phenomena and auditory ones in particular are located so to say on a smooth continuum may seem to be at odds with contemporary theories of categorical perception (see Harnad 1987).[3] Differences (and similarities) while they can be perceived as gradual (and hence, so to say, as quantitative matter) as is the case for shades of colors, can also be perceived as more sharp or blunt (and as a matter of quality). Perceptions of the first type are called continuous, perception of the second type categorical. One could thus object that Mehlberg's ontology of the mind is inaccurate or at best incomplete. Contrary to what this objection presupposes, Mehlberg is not, however, necessarily committed to the view that *all* perceptions are continuous but only to the view that those that concern "intrinsic" features of objects are. In the light of examples of such intrinsic features which he gives (e.g., volume, timbre, pitch, duration), he seems to be right about the continuous nature of the perceptions involved. So, for the sake of argument, and because what he says on the topic leads him to results that are otherwise intellectually engaging, let us concede the point.

---

[3] We'd like to thank an anonymous referee for the useful suggestion that led us to make this point more precise.

As we've just mentioned, Mehlberg believes that our perceptions of intrinsic features of internal, mental states is gradational. His point, and the one which needs be emphasized here is the following:

> if two mental phenomena share all gradational properties, then they do not differ with respect to their sensory, convictional and volitional components, and, therefore, they are (...) intrinsically similar (1980a, 43).

Mehlberg goes on to argue that an adequate characterization of the notion of intrinsic similarity requires additional precision: in order to be intrinsically similar objects must share more precisely the same "shades" of these gradational properties. According to Mehlberg, the shade of a gradational property is to be defined as follows:

The quality q is a shade of the gradational property P whenever the ensuing two conditions are satisfied:

(a) two entities *E, E'* sharing the quality q do not differ from one another with regard to P
(b) if an entity E possessing the quality q does not differ from another entity *E'* with regard to the corresponding property P, then the latter entity (i.e. *E'*) also has the quality q.

In this sense, a *C major* is a shade q of the pitch P. Take two notes*E and E'* that are of that shade, one played on a piano, the other sung by a solfege student. *E and E'* do not differ with regard to their pitch P (although they may differ with respect to their timbre *P'*, for instance). In replicating adequately the note E that her teacher has played on the piano, the singer has produced a note *E*' that has the same shade. The notion of shade is interesting inasmuch as it allows Mehlberg to introduce something between the type to which an object belongs and a token of that object. An object is an instance of a gradational property P by virtue of its being an instance of a shade q of that property. We will see that the distinction plays an important role in Mehlberg's account of multiple realisation.

One tentative reformulation of the correspondence thesis Mehlberg immediately rejects is the following:

(T1b) Whenever two states of a nervous system are of the same shade in all respects, then their accompanying experiences are likewise of the same shade in all respects. (Cf. Mehlberg 1980a, 45; cf. also: 1980b, 274).

As Mehlberg poins out, it is extremely improbable that two states of a system as complicated as the human nervous system could ever be of the same shade in all respects so that one could be an exact replica of the other (Mehlberg, *Parallelism*, 45). If the nervous system is indeed too complex to ever instantiate states that are identical in the latter sense, then (T1b) can never be fulfilled. Mehlberg consequently weakens the condition: the correspondence is not between classes of mental states, that are intrinsically similar on the one hand, and classes of brain states that are intrinsically similar, on the other. The correspondence holds between classes of mental states that are (merely) almost (intrinsically) similar and classes of brain states that are almost (intrinsically) similar. In other terms, the correspondence holds between individuals of a certain type, where the type is not indeed defined by identity but by mere similarity – or what is perhaps closer to Mehlberg's terminology: quasi-similarity.

What about correspondence itself? Although this is questionable, Mehlberg thought that the correspondence between tokens of (certain types of quasi-similar) mental states and tokens of (certain types of quasi-similar) brain states could be explained by means of the mathematical concept of a continuous function. (Mehlberg, 1980a, 46, Mehlberg, 1980b, 275), where the definition of a continuous function he has in mind is the following: a function $f\ (x)$ is continuous at the point x_0 if f assigns to values of x sufficiently close to x_0 arbitrarily close values of $f\ (x)$ (Mehlberg 1980b, 274). Mehlberg probably thought that resorting to the concept of a continuous function could help him, not because of some technical feature of the mathematical definition of continuity itself, but because the general utility of the notion function when it comes to mapping values onto other values and because of the reference to the idea of a "sufficiently close value" of x which can be seen as epitomizing the idea behind Mehlberg's notion of quasi-similarity: in order to give an output "sufficiently close" to a value or of a certain type defined by quasi-similarity, the input need only be "sufficiently close" to a certain value, i.e. belong to a type defined by quasi-similarity.

On Mehlberg's interpretation the correspondence thesis holds that there are functions whose arguments are tokens of shades of brain states of a certain type (shades of a certain neurophysical property) and whose corresponding values are tokens of shades of mental states of a certain type (certain mental properties). These functions, and this is what is relevant, have as a range of arguments a set of shades that are all shades of the same neurophysical property and as a range of value, a set of quasi-similar shades of the same mental property (Cf. Mehlberg 1980a, 45). Mehlberg seems to have something of the following sort in mind: when our nervous system – in the previous example, the auditory system – reacts appropriately, i.e. when our brain is caused by the external world to instantiate a given shade of a neurophysical property, we simultaneously have given auditory experiences that instantiate shades of some mental properties. For the resolution of the problem of multiple realization, this distinction between traditional mental property token and the less orthodox token of a shade of a certain property comes in quite handy. Although he does not do so, Mehlberg is in a position to claim that both I and a cat can feel pain although we do not have exactly the same neurological equipment and although the pain we feel may be qualitatively different. What we experience are tokens of quasi-similar shades of the same mental property which corresponds – Mehlberg says "is determined" – by tokens of quasi-similar shades of a certain neurophysical property. There are indeed common features to human brains and cat or mice brains by virtue of which they could be said to be quasi-similar: it is those common features that make it possible to make previsions about the effect of, for instance, certain drugs on levels of neurotransmitters or hormones in the human brain based on research conducted on animals. Likewise, Mehlberg's account can explain why slight variations in one's neurophysiological states do not necessarily imply that one suddenly finds oneself in another mental states. As long as the states I am in instantiate quasi-similar shades, the type of mental state in which I find myself remains the same.

Now, according to Mehlberg, there are two kinds of properties, micropsychophysical and macropsychophysical properties. There are consequently two corresponding kinds of "parallelism" or epiphenomenalism: microepiphenomenalism which takes as input physical properties of the microscopic level and macroepiphenomenalism which takes as input physical properties of the macroscopic level. Mehlberg does not provide us with concrete examples but the latter would not be difficult to find. Besides, what Mehlberg is primarily interested in is a more general discussion of the relation between the microscopic and macroscopic levels.

What is the relationship between these two kinds of properties? For one thing, according to Mehlberg, whilst microphysical properties are "primary", macrophysical properties are "derivative". Mehlberg gives the following definition of a derivative property:

(Derivative) A property *P* is derivative with regard to a property *P'* if the shade of *P* attributable to a given phenomenon is a continuous function of the shade of *P'* ascribable to the same phenomenon.

When this is the case, the "domain" – a set of shades with at least two elements *q* and *q'* constitutes a domain of shades if it includes all shades which lie between *q* and *q'* – of shades of *P'* always corresponds to the same shade of *P* (Cf. Mehlberg 1980a, 46; 1980b, 276).

(Primary) A gradational property *P* is primary if it is not derivative with regard to any gradational property (Cf. Mehlberg 1980a, 46; 1980b, 276).

According to Mehlberg, "derivative properties are exemplified by statistical properties which depend upon some averaging" (Cf. Mehlberg 1980a, 46; 1980b, 276). A derivative property has therefore a certain extension of variability. This explains, in Mehlberg's eyes, why a melodic pattern of sound configurations is the same despite changes in the elements of the configuration. Imagine for instance a musical competition where musicians are asked to perform the same piece. One's small mistake in performing, caused, for example, by replacing a quaver with a semiquaver changes the configuration (often the changes of configurations are more important) but this does not imply a difference in the melodic pattern. This said, in contemporary research on musical perception researchers emphasize the difference between music perception by musicians and non-musicians. Musicians (especially those who have absolute hearing) are able to notice changes that non-musicians fail to notice. We leave here open the consequences of these results for Mehlberg's thesis.

Mehlberg's distinction between micro- and macroepiphenomenalism is based on the distinction between primary and derivative properties:

(Microepiphenomenalism) The properties of mental phenomena (experiences) are determined by *primary* properties of their substrata, i.e. the brain states to which they correspond.

(Macroepiphenomenalism). The properties of mental phenomena are determined by certain *derivative* properties of their substrata. (Mehlberg, *Parallelism*, 46).

Interestingly, Mehlberg sets forth arguments against both kinds of epiphenomenalism. Mehlberg disagrees with the idea that correspondence between the mental and the physical is governed by "deterministic" laws. Instead, he defends what can be termed "statistical epiphenomenalism", a view which rest on the idea that the correspondence between brain states and mental states – the same holds for pretty much all regularities that characterise living organisms in general – are governed, not by deterministic but by statistical laws. This view, however need not be discussed here.

The kind of psychophysical parallelism Mehlberg criticizes and which leads him to his own epiphenomenalistic account of the mind is rooted in dualistic assumptions that philosophers from Twardowski's School would not have been not unlikely to have made. Twardowski's students developed a philosophy of mind that had its source in Brentano's early descriptive psychology. Twardowski and most of his students distinguished between mental and physical phenomena. They considered the mental to be non eliminable and the problem of explaining subjective sensations – and more generally the Cartesian problem of the self – played no small role in their view. Twardowski tackles this question in a polemic with Fechner that was popular among psychologists of the time (1895). In this Brentanian tradition, Tadeusz Kotarbiński's "*pansomatism*" (a form of reism Kotarbiński developed in the 1930's) is somewhat of an oddity. It was, at any rate, the only truly monistic approach to the mind body problem that was defended in Poland. Pansomatism is a form of ontological physicalism (or materialistic monism) that implies that "the soul" (the traditional *psyché*) is no more than the brain (or a nervous system or something physical in general).[4] In "On the Essence of Inner Experience" (1922) Kotarbiński had provided an analysis of statements of everyday language in which there are psychological terms.[5] Although Kotarbiński's approach can be seen as an interesting alternative to behaviorism (See Woleński 1989), it remained, like a number of other ideas developed by Twardowski's pupils including but certainly not limited to Mehlberg's unknown to the rest of the philosophical world. Nonetheless, just like Mehlberg, Kotarbiński developed systematic considerations that formed a unified philosophical system. He combined a radical realist ontology with a realist epistemology. In this light, what both philosophers have in common with nowadays philosophy of mind is the scientific approach that was found for the first time in their work and a carefully analytic methodology that is characteristic of Twardowski's entire school.[6]

---

[4] See (Woleński 1989), especially: 229–232.

[5] Woleński compares Kotarbiński's analysis to Ryle's analysis from 1930 (Woleński 1989, 231).

[6] This is a vastly reworked version of a paper that was originally presented at the Colloque *Le rayonnement de la philosophie polonaise au XXe. L'héritage philosophique de Kazimierz Twardowski*, November 2004, in Paris I would like to thank Victor Rosenthal who was the commentator of my paper, the anonymous referee, and especially Sandra Lapointe for her effort to make this text much clearer.

Bibliography

Harnad, Stevan (ed.) (1987), 'Psychophysical and cognitive aspects of categorical perception. A critical overview'. In *Categorical Perception: The Groundwork of Cognition*, New York, Cambridge University Press.

Kotarbiński, Tadeusz (1922), 'O Istocie Doświadczenia Wewnętrznego' (On the Essence of Inner Experience), *Przegląd Filozoficzny* 23, 149–170.

Kotarbiński, Tadeusz (1935), 'Zasadnicze Myśli Pansomatyzmu', *Przegląd Filozoficzny* 38. Eng. trans. A. Tarski and D. Rynin, 'The Fundamental Ideas of Pansomatism', *Mind* 64 (1955).

Mehlberg, Henryk (1980a), 'The Present Empirical Status of Psychophysical Parallelism. Supplement', In *Time, Causality and Quantum Theory. Studies in Philosophy of Science*, R. S. Cohen (ed.), Dordrecht, Reidel, 261–268.

Mehlberg, Henryk (1980b), 'Conceptual Analysis of Psychophysical Parallelism. Supplement', in: H. Mehlberg, *Time, Causality and Quantum Theory. Studies in Philosophy of Science*, R. S. Cohen (ed.), Dordrecht, Reidel, 269–284.

Twardowski, Kazimierz (1895), 'Metafizyka Duszy' (Metaphysics of the Soul), *Przełom* 3, Vienna, 467–480.

Woleński, Jan (1989), *Logic and Philosophy in the Lvov-Warsaw School*, Dordrecht, Kluwer.

Chapter 12
Leopold Blaustein's Analytical Phenomenology

Wioletta Miskiewicz

There are at least two kinds of phenomenology: hermeneutic and descriptive. The latter rests on the idea that what is given in conscious experience is direct, akin to perception – what most phenomenologist would have called "intuition" – and is therefore a genuine source of knowledge. The theories of early *analytical* phenomenologists were aimed at providing an understanding of the latter. For instance, the well known distinction between the quality, the content, and the object of mental acts elaborated by Twardowski, Husserl and Meinong in the wake of Brentano was meant as a conceptual tool for the purpose of analyzing and describing cognitive processes such as "representation" and "judgment". Leopold Blaustein, a Lvovian philosopher who belongs to Twardowski's School, sought to develop the full heuristic value of this theory of intentionality and, for one thing, his results can be seen to have anticipated some of the most significant aspects of contemporary theories of cognition, and of perception in particular.

Leopold Blaustein was born in 1905. He studied philosophy and German philology in Lvov and was part of the last generation of Twardowski's students. His private and academic correspondence with Twardowski, as well as the fact that he dedicated most of his books to him, show that Blaustein remained close to the founder of the Polish Analytical School throughout his life. Blaustein also attended the lectures of Ajdukiewicz and Ingarden – both students of Twardowski's. Husserl had a very significant impact on him: in 1925, while working under Twardowski on a thesis about Husserl's theory of objects, Blaustein went to Freiburg im Breisgau. There, he was deeply impressed by Husserl, both as a philosopher and as a person. While he admired Husserl's philosophical commitment, Blaustein was however skeptical about the kind of philosophy Husserl was developing.[1] The reason for this will become clearer in what follows.

W. Miskiewicz (✉)
IHPST (CNRS /Paris1/ ENS)/Archiwum Kazimierza Twardowskiego, Warsaw, Poland
e-mail: wioletta.miskiewicz@ens.fr

[1] Blaustein published his recollections of the stay in Freiburg in (1930).

S. Lapointe et al. (eds.), *The Golden Age of Polish Philosophy*, Logic, Epistemology, and the Unity of Science 16, DOI 10.1007/978-90-481-2401-5_13,

In 1927 Blaustein defended his thesis: "On Husserl's theory of act, content and object of presentations" and obtained his Ph. D. He passed the *rigorosum* examinations and, while continuing his research, started working as a teacher of German and of philosophy in a high school in Lvov. At the end of his first year as a teacher, he received a scholarship for a study trip to Germany.[2] Blaustein moved to Berlin where he frequented Köhler's Institute of Psychology and where the atmosphere, was favorable to his intellectual development. During his stay he studied *Gestalt* psychology, investigating in particular the technical aspects of experimental research. After Blaustein went back to Poland, he resumed his work at the *Gymnasium*, but kept an active interest for research in psychology, pedagogic and media communication. As a secular Jew, he also had an interest in questions relating to the Jewish community in Poland. He married Eugenia Ginsberg[3], a pupil of Ajdukiewicz, and they had one child. The exact circumstances of his death are not known: he may have been executed by the Germans in the Lvov ghetto in 1942 together with his wife and their little boy. According to other sources, he committed suicide later, in 1943 or 1944.

In his death, Blaustein shares the destiny of many of Twardowski's students[4], and he might have been one of the most remarkable of them. Judging by their writings, he and his friend, Walter Auerbach[5] were apparently developing an entirely new branch of phenomenology. Their programme was analytic, descriptive and interdisciplinary. Their phenomenology was deeply related to experimental psychology. That one of Twardowski's students should have had interests in the latter should not be surprising. Twardowski had been the founder of the first laboratory of experimental psychology in Central Europe, a laboratory in which doctor degrees in psychology were delivered as soon as 1898, and which produced a large number of psychologists (such as Witwicki, Baley, Kreutz, Błachowski).

One can roughly distinguish two trends in Blaustein's work. On the one hand, the *theoretical* work focuses on the problem of intuition in the sense of immediate, evident knowledge, as well as on the typologyof mental states[6]. On the other hand, he published on *applied* topics, in the philosophy of arts[7] and the philosophy of media[8], and on questions of education[9] – liberally commenting, among other things,

---

[2] The exchange with Twardowski that dates from this period is interesting in more than one respect: in addition to allowing us to gain an insight of Blaustein's complexe personality, it contains a first-hand description of Berlin's academic life and of important figures of the time such as Stumpf, Baumgardt or Wertheimer. See Jadczak 1993.

[3] See Ginsberg 1929, 1938.

[4] Jadacki and Markiewicz (eds.) 1993 contains 36 biographies of philosophers who belonged to Twardowski's circle and who were Jews or of Jewish ascendance.

[5] See Auerbach 1931, 1935. Walter Auerbach (1900–1942) was a student of Twardowski, Ingarden and Kotarbiński.

[6] See Blaustein 1928, 1930a, b, 1931a, b.

[7] See Blaustein 1937.

[8] See Blaustein 1939, 1933.

[9] See, for instance, Blaustein 1934.

on the laziness and the lack of discipline among high school pupils. In spite of his early death (he was less that forty years old), Blaustein left more than fifty articles and books.

In the German-speaking literature, the question of immediate knowledge is often dealt with under the heading of *Anschauung*. For this reason, translating *Anschauung* as *intuition* is problematic. At any rate, it is this question, the question of immediate knowledge or *naoczność* or *Anschauung*, which constitutes the core of Blaustein's theoretical work.[10] The question of the immediate evidence of knowledge is at the origin of his typology of mental states and of the analytical apparatus which is the basis of his approach. The problems which Blaustein studied have their roots directly in Twardowski's thought, and in particular in the distinctions Twardowski introduces in the published version of his habilitation thesis (*On the content and Object of Representation,* Wien 1894), between direct and indirect knowledge.

Twardowski believed that it is impossible to achieve a correct classification of mental states without taking into account the Kantian distinction between *Anschauungen* (*concrete, intuitive and direct* presentations – Kant's intuitions) and *Begriffe* (*abstract, non intuitive and indirect presentation* – Kant's concepts). As a matter of convenience, Twardowski even when writing in Polish used the German terms.[11] What's relevant here is the following. For Twardowski, evidential cognition is a discriminatory criterion when it comes to determining, among representations, the ones that are "direct" and "intuitive" – we'll follow the traditional terminology but it is important to note that what Twardowski has in mind when he speaks of intuition is something akin to or that implies perception – as opposed to those that are "indirect" and "conceptual". In this respect, what's most interesting about Blaustein's theory is the fact that his theory of direct presentations is based on a development that can clearly be traced back to Brentano's original theory. Blaustein, following Brentano, adopts the distinction between the matter and the quality of the mental act. But, in turn, as a pupil of Twardowski, he distinguishes within the matter of the act between the content (*Inhalt/Gehalt*) and the object (*Gegenstand*).[12] What is more, Blaustein adopts the Husserlian distinction, within the matter itself, between the intuitive content – which corresponds roughly to what we usually call sensation – and the objective meaning (*Bedeutung*). In his doctoral thesis, Blaustein further argues for the necessity of attributing a "modality" to sensation. In his view, then, the matter

[10] See Chudy 1981; Debowski 1996.

[11] Cf. Twardowski 1894, 109. Although this distinction is presupposed by the entire treatise, it is made explicit only in the last part of the last paragraph where Twardowski mentions Kant and quotes Riehl: "Der Anschauung als der unmittelbaren Vorstellung eines Gegenstandes steht der Begriff als dessen mittelbare gegenüber, als Vostellung desselben durch andere Vorstellungen oder einen Teil der anschaulichen Gesammtvorstellung."

[12] Cf. Woleński 1999, 35. There are some doubts concerning Twardowski's understanding of objects. As Blaustein remarks, there are two conflicting views in Twardowski: (a) objects are items to which mental acts are directed, independently of their ontological status (real, possible, etc.), and (b) objects are real phenomena (Blaustein 1928, 17).

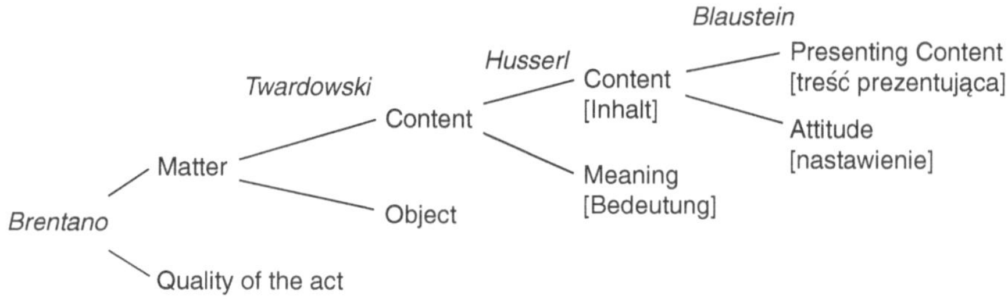

Fig. 12.1 Genealogy of Blaustein's "grasping attitude"

of a mental act is the (non-separable) part of the act that determines the direction of the intention: the act is directed towards this or that object. But while the matter of the act specifies the sensorial qualities of objects, it also always does so in a certainmode: "for real" (in which case what we have is a *bona fide* perception of actually, materially existing objects), or "as if" (in which case, what we have is, for instance, an act of imagination). Blaustein's original addition thus consists in enriching Husserl's conception of intuitive content by introducing an aspect pertaining to the *modality* of the act of presentation (or imagination, memory etc). As we will see in what follows, Blaustein describes this modality as arising from the "grasping attitude" of the agent. Blaustein claims that in order to account adequately for the *intuitive character* of objective data, one must distinguish within Husserl's notion of content, in addition to the act matter, the "grasping attitude", which is an equally determining factor in the perception of the intentional object. Blaustein distinguishes the "grasping attitude" from the purely intuitive or purely sensorial "presenting content" of sensation[13]. Now, what is interesting is that according to Blaustein, the "grasping attitude", inasmuch as it is an aspect of sensation, pertains just like the latter to the properties of the object grasped. The latter however he conceives not as a phenomenon, but as an object being part of the natural world, e.g. material or virtual. We see here, in Ingarden's follower, the same commitment to realism concerning the existence of the external world and a similar aversion to idealism.

We usually use the term, 'modality' to refer to the way in which judgments are true or false. Following Blaustein, I use the term somewhat differently. I call "modality of the matter of a mental act" the *way* in which the matter (see Fig. 12.1) of an act specifies the qualities of objects. The modality of an intentional act refers to the ontological status of the object (actual, imagined, schematic, symbolic) and the latter, as we have seen, depends on the "grasping attitude". Blaustein hence distinguishes different types of matter.[14] Now, what is interesting and truly original about Blaustein is his observation that whether an object, or one of its determinations, is effective or fictive, for instance, is a function of the way in which the matter of

[13] See Blaustein, 1928, 78; Auerbach 1930, 210–211, 1931, 214–216.

[14] Blaustein distinguishes different types of matter that are determined by the way in which one refers to the intentional object (perception, judgment, will, feeling, imagination, memory, *etc.*)

the act specifies the qualities of the object, that is, it is a function of the "grasping attitude".

Blaustein's move raises a number of questions. What role does the distinction between the presenting content and the modality of the matter of an act plays in an adequate account of obvious, immediate knowledge? Is Blaustein's solution, if he has one, formally coherent? Does he avoid, unlike Ingarden, the old alternative regarding the status of sensual data in the presentations, namely that sensation is either already an interpretation, or that it possesses the quality of what it represents (the question as to whether the vision of a red coat is itself red)?

In his doctoral thesis Blaustein had tried to understand the way in which sensorial content is part of direct presentations. Do they merely accompany direct presentations, or are they effective elements of the latter? Can a quality like "real" or "imaginary" be spotted at the interface of consciousness and the real world? Husserl and Ingarden had tackled this question and although there is no space here to present the interesting exchange between them regarding the immediacy of sensual data in intentional acts[15], it must be stressed that it constituted the most technical aspect of their controversy concerning idealism. For Blaustein this problem, which is connected with issues concerning the different degrees of evidence of mental acts, was also the narrow door to a solution to the problem of the possibility of immediate knowledge.

Blaustein, on the basis of a distinction between different degrees of evidential cognition, identifies seven types of direct presentations (as opposed to concepts): (i) sensations proper, (ii) perceptive presentations, (iii) reproductive presentations, (iv) creative presentations, (v) imaginary presentations, (vi) schematic representations, (vii) meaningful representations. He analyses the latter in a series of works in which he progressively moved away from Brentano. He came to the conclusion that the Brentanian distinction between the quality and the matter of the act was the result of an abstraction and was therefore not self-evident. Against Twardowski, Husserl and Meinong who defended its relevance, he considered Brentano's distinction to be merely tentative. Blaustein's alternative conception of the structure of the mental act and, in particular, his notion of the content's sensorial modality (the grasping attitude) allows him to argue, against Brentano, that "imaginary presentations of imaginary objects cannot be the psychological basis of acts of judgment in which the object would be considered as actually existing" since the latter is already "a specific modification of the way in which the matter of the act attributes its qualities to the intentional object."[16] While the matter of an act of effective perception gives the qualities to its object "for real", the matter of an imaginary act attributes them as 'quasi'-association.

Hence, in summary, according to Blaustein, the modality of an act is a property of the matter of that act. It is thus specific to sensation and relates directly to the object, which exists independently. Nonetheless, its also has its origin in a given attitude of

---

[15] See Miskiewicz 2003a.

[16] See Blaustein, 1928, 27.

she who actually perceives. The relation between these two aspects is complex and is perhaps best illustrated in Blaustein's philosophical aesthetics. Blaustein is not interested in the work of art as an object of beauty, but rather in the way in which the aesthetic agent reacts in presence of a work of art. For Blaustein, perceiving an object is always *observing an object with a certain attitude* and he sought to study the role of the act of perceiving in one's reception of aesthetic artefacts. In this respect, Blaustein's understanding of perception however differs from that of classical phenomenology. This is why Blaustein chooses to use somewhat awkwardly a new term, foreign to the traditional phenomenological jargon, namely 'observation', to designate the manifold of sensations caused by the presence of the work of art: sensual, emotional and intellectual impressions.

While Blaustein aims at understanding what type of *modification* is involved in the sensations that accompany the perception of artworks, what he says holds indeed for objects in general. He seeks to demonstrate that sensations cannot be considered to be the only basic sensorial data of aesthetic experience. Sensorial data, in themselves, lack any traces of cultural performances, and as such are objective in the sense that they are the same for everyone (they are intersubjective). But, according to Blaustein, the agent always actively *structures* the process of sensation. Let us consider a figurative painting, the Mona Lisa, for instance. Let us imagine that we're in the presence of this painting in the Louvre. We can perceive it either as a heavy golden frame and an old canvas covered with different kinds of colouring agents, or we can – in a somewhat different attitude – look straight at the face which is represented, stroll the soft landscape sketched behind it. We can also leave Mona Lisa for a while and reflect on the security measures taken to protect the famous object against possible thieves and despoilers, and then, after briefly returning to the actual painting, ask ourselves whether the protecting glass doesn't alter the precision and the pleasure we take in Da Vinci's work, etc. What Blaustein wants to point out is the fact that while we have the same Mona Lisa-induced sensory impressions in all these different scenario, the ways in which the latter present themselves to us differ widely. For Blaustein this difference reveals an important fact about perception and representation in general. It is precisely in order to account for this fact that Blaustein introduces the distinction between the presenting content and the grasping attitude. The presenting content is a sensory complex which – together with "*Gestalt* qualities" – makes the object available to consciousness. The same presenting content however can be subject to different modalities.

It is hardly impossible not to think at this point of Husserl's so-called "natural attitude" and to ask how the latter compares to Blaustein's "grasping attitude". What is striking is that if Blaustein is right, that is, if the matter of a mental act already contains data relative to the mode of existence of the object then, firstly, contrary to what Husserl assumes, the suspension of the natural attitude is not unproblematic and, secondly, for this very reason, naturalisation[17] could be the unavoidable upshot of all phenomenological analysis.

[17] *See* Miskiewicz 2003b.

Kotarbiński described Leopold Blaustein as a "reliable and obstinate man"[18] and Ingarden added that he "distinguished himself by subtle, penetrating intelligence, by a broad spectrum of theoretical interests and by his considerable capacity for work". But Blaustein was more than that.[19] For he strikes us as well by the great creativity, originality and daring thought he expressed clearly and illustrated with sobriety. Blaustein anticipated (with great subtlety) many theories, such as Wollheim's "seeing-in" in aesthetic theory[20] and McLuhan's "the medium is the message"[21] well know in media study to mention only two[22].

Blaustein stands in the direct prestigious lineage of Brentano, Twardowski and Husserl. We showed that he introduced a significant development within this tradition. But, in conclusion, we can mention another aspect, clearly modern, of his work: its interdisciplinarity. And, it should be stressed that the latter is not ideological interdisciplinarity. In his *Study on the perception of radiophonic plays* Blaustein writes: "It is not possible not to overstep the boundary of psychological enquiry"[23] and he then bases himself on two sciences: psychology and aesthetics. Things by their very nature necessarily direct the researcher from one science to the other.

Bibliography

Auerbach, W. (1931) *O wątpieniu* (On Doubt), Lvov, Księga Pamiątkowa PTF.

Auerbach, W. (1930), 'Review of Husserlowska nauka o akcie, treści i przedmiocie przedstawienia' (Husserl's theory on act, content and object of representations), *Przegląd Humanistyczny* 5/2, 210–211.

Auerbach, W. (1931), 'Review of Przedstawienia imaginatywne' (Imaginative Presentations), Przegląd Humanistyczny, 6/1–2, 214–216.

Auerbach, W. (1935) 'Zagadnienie wartości poznawczej sądów przypomnieniowych' (The Question of the Heuristic Value of Memory Judgments), *Kwartalnik Psychologiczny* 7, 25–60.

Blaustein, L. (1928), *Husserlowska nauka o akcie, treści i przedmiocie przedstawień* (Husserl's theory on act, content and object of representations), Lvov, Archiwum Towarzystwa Naukowego we Lwowie.

Blaustein, L. (1930a), 'Edmund Husserl i jego fenomenologia' (Edmund Husserl and his Phenomenology), *Przegląd Humanistyczny* 2. See also 'Próba krytycznej oceny fenomenologii' (Essay of a Critic of Phenomenology), *Ruch Filozoficzny* 11, 1928/9, 164–166.

Blaustein, L. (1930b), *Przedstawienia imaginatywne. Studyum z pogranicza psychologii i estetyki* (Imaginative Presentations), Lvov, nakładem Polskiego Towarzystwa Filozoficznego.

Blaustein, L. (1931a), *Przedstawienia schematyczne i symboliczne. Badania z pogranicza psychologji i estetyki* (Schematic and symbolic representations. Studies on the border of psychology and aesthetics), Lvov, nakładem Przeglądu Humanistycznego.

[18] Jadacki and Markiewicz, 1993, 11.

[19] See Ingarden 1946, 335–336; Gromska 1948, 63–67; Domagała 1990, 125–128.

[20] See Wollheim, 1980, 1988; See also Scruton, 1982; Lopes, 1999.

[21] See. McLuhan, 1964.

[22] See Rosińska 2001, 2005.

[23] *Blaustein 1939*, 6.

Blaustein, L. (1931b), 'O naoczności jako właściwości niektórych przedstawień' (On intuition as a property of some representations), Lvov, Księga Pamiątkowa Polskiego Towarzystwa Filozoficznego we Lwowie.
Blaustein, L. (1933), 'Przyczynki do psychologii widza kinowego' (On the psychology of a moviegoer), *Kwartalnik Psychologiczny* 4, 192–236
Blaustein, L. (1934), 'O lekturze filozoficznej w szkole średniej' (On Reading Philosophy in Colleges), Lvov, Gimnazjum 2.
Blaustein, L. (1937), 'Der imaginative Kunstgegenstand und seine Gegebenheitsweise' (The Imaginary Work of Art and Its Mode of Givenness), *Deuxième Congrès International d'Esthétique et de Science de l'Art*, Paris, Alcan, 245–249.
Blaustein, L. (1939), 'Studies on the perception of radiophonic pieces', *Kwartalnik Psychologiczny* 11, 106–161
Chudy, W. (1981), 'Zagadnienie naoczności aktów poznawczych' (The Problem of the Evidence of Cognitive Acts), *Roczniki Filozoficzne* 1, 185–194.
Dębowski, J. (1996), 'Naoczność i jej rodzaje w ujęciu K. Twardowskiego i L. Blausteina' (Evidence and Its Kinds in the Conceptions of Kazimierz Twardowski and Leopold Blaustein), *Annales Universitis Mariae Curie-Sklodowska* 21–22, 75–96.
Domagała, T. (1990), 'Leopold Blaustein – nauczyciel folozofii', *Edukacja Folozoficzna* 10, 125–128.
Ginsberg, E. (1929), 'Zur Husserlschen Lehre von Ganzen und Teilen', *Archiv für systematische Philosophie* 32, 109–120.
Ginsberg, E. (1938), 'W sprawie pojęć samodzielności i niesamodzielności', Księga pamiątkowa PTF we Lwowie, Lwów, Książnica Atlas, 143-168; Engl. Trans. 'On the concept of existential dependence and independence' in Smith, B. (ed.) *Parts and Moments. Studies in Logic and formal Ontology*, Munich, Philosophia, 1982, 266–288.
Gromska, D. (1948), 'Philosophes polonais morts entre 1939 et 1945. Leopold Blaustein', *Studia Philosophica* 3, 63–67.
Ingarden, R. (1946), 'Leopold Blaustein', *Przegląd Filozoficzny* 42, 335–336.
Jadacki, J. and Markiewicz, M. (eds.) (1993), *...A mądrości nic nie przemoże*, (Wisdom Will Never Be Defeated), Warsaw, Polskie Towarzystwo Filozoficzne.
Jadczak, R. (1993), 'Uczeń i nauczyciel. Z listów Leopolda Blausteina do Kazimierza Twardowskiego z lat 1927–1930' (Disciple and Master. The Correspondence Between Leopold Blaustein and Kazimierz Twardowski (1927–1930)), in Jadacki and Markiewicz 1993.
Lopes, D. (1999) *Understanding Pictures*, Oxford, OUP.
McLuhan, M. (1964) *Understanding Media: The Extensions of Man*, New York McGraw Hill.
Miskiewicz, W. (2003a), 'Ingarden et la réception de Brentano en Pologne', *Les Etudes Philosophiques*, 2003/1, 83–97.
Miskiewicz, W. (2003b), 'Husserl contre psychologisme et cognitivisme: la naturalisation superficielle et la psychologie d'un style nouveau dans les Recherches Logiques', *Aux origines de la phénoménologie*, D. Fisette and S. Lapointe (eds.), Paris/Quebec, Vrin/PUL, 221–240.
Rosińska, Z. (2001), *Blaustein. Koncepcja odbioru mediów.* (Blaustein. A Conception of Media), Warsaw, Prószyński i S-ka.
Rosińska, Z. (2005), *Leopold Blaustein. Wybór Pism Estetycznych* (Leopold Blaustein. Selected Papers in Aesthetics), Cracow, Universitas.
Scruton, R. (1982), *Art and Imagination*, London, Routledge and Kegan Paul.
Twardowski, K. (1894), *Zur Lehre vom Inhalt und Gegenstand der Vorstellungen: Eine psychologische Untersuchung*, Vienna, Alfred Hölder. Eng. Trans. *On the Content and Object of Presentations*, The Hague, Martinus Nijhoff, 1977.
Woleński, J. (1999), 'Twardowski and the Distinction between Content and Object', *Brentano Studien* 8 (1998/99), 15–35; reprinted in J. Woleński, *Essays in the History of Logic and Logical Philosophy*, Cracow, Jagiellonian University Press, 22–35.
Wollheim, R. (1980), *Art and Objects*, Cambridge, CUP.
Wollheim, R. (1988) *Painting as an Art*, London, Thames and Hudson.

Part IV
Around Twardowski's School

Chapter 13
Nonclassical Conceptions of Truth in Polish Philosophy at the Beginning of the 20th Century

Katarzyna Kijania-Placek

13.1 Introduction

This paper is about nonclassical theories of truth in Poland at the beginning of the 20th century, both in the Lvov-Warsaw School and outside of it. I shall concentrate on two philosophical approaches which in my opinion are interesting but not commonly known most likely because the relevant papers were written only in Polish: Władysław Biegański's, whose theory is based on biological evolutionism, and Edward Poznański and Aleksander Wundheiler's theory of truth in physics. My paper is mainly historical and my aim is to nuance the picture according to which Polish theories of truth are almost invariably exclusively associated with the views of Tarski and his teachers. I agree with Simons and Woleński[1] that Poles were obsessed with truth but I hope to show that this obsession took many forms.

13.2 Biegański

Biegański (1857–1917) did not have an academic career and he was not a professional philosopher. He was a medical doctor in Częstochowa, a provincial town to the north of Krakow. He nonetheless wrote many books and papers, mostly in the philosophy of medicine. He was an important member of the second generation of the Polish School of philosophy of medicine founded in Warsaw by Tytus Chałubiński in the second half of the nineteenth century.

Biegański's papers in the methodology of medicine were internationally acknowledged, especially after the second edition of his *Logic of medicine* was published in 1908 (first edition 1894). The book was translated into German in 1909 and was

---

K. Kijania-Placek (✉)
Jagiellonian University, Cracow, Poland
e-mail: uzkijani@cyf-kr.edu.pl

The research for this paper was supported by the Foundation for Polish Science. Parts of the content of Section 3 are based on my book (Kijania-Placek, 2000).

[1] Woleński and Simons, 1989.

S. Lapointe et al. (eds.), *The Golden Age of Polish Philosophy*, Logic, Epistemology, and the Unity of Science 16, DOI 10.1007/978-90-481-2401-5_14,

reviewed more than 50 times in medical journals in German, English and French. In addition to papers in medicine and the philosophy of medicine, Biegański wrote several logic textbooks, two books on ethics and two books on epistemology: *Treatise on cognition and truth*, published in 1910 and *Epistemology from the teleological standpoint* in 1915. Although Biegański can hardly be considered to be a member of the Lvov-Warsaw School, he was familiar with Twardowski's writings and, in particular, as we will see in what follows, with Twardowski's paper "On so-called relative truths". Moreover, Biegański and Twardowski were both on the editorial board of the *Medical Weekly*, published in Lvov after 1907.

Biegański presented his conception of truth in the *Treatise on cognition and truth*, published in Warsaw in 1910. In this book he criticises what he calls the "classical" conceptions of truth and knowledge. His conception of truth is based on his theory of cognition, which he discusses in the first part of the book. Biegański develops a theory of truth, according to which truth proper and falsity proper are not properties of judgments, but are themselves judgments asserting justification (or lack of justification) of other judgments (1910, 147). Truth is not a property of a judgment in the sense that it is not determined by the content or form of the judgment. Judgments, if considered independently of the knowing subject, are neutral with respect to truth or falsity. In the same sense, impressions are pleasant or unpleasant only with respect to the experiencing subject (1910, 134, 136). What makes judgments true or false is their comparison with something else. According to Biegański, this comparison is not a comparison with some external reality, but with knowledge we have previously acquired (1910, 134–135, 139).[2] A judgment is true if it is completely justified by the principles from which it is inferred.

> [T]ruth can be defined as a judgment about another judgment, that the content of the latter is sufficiently justified, i.e. that it agrees with the principles from which the judgment has been inferred (1910, 150).[3]

By 'principles' Biegański means impressions or other judgments, upon which the justification of the judgment in question is based. Consider the judgment:

Murder is outrageous.

Whether this judgment is true can be determined by considering our concepts and judgments about morality, other human beings, community life and its obligations, etc. The judgment is true as long as it is justified by these principles. For that to be

---

[2] At some point Biegański even claims that " 'true' and 'false' are reactions of a cognitive mind to judgments and are states of consciousness that accompany judgments" (1910, 134–135). Because I think that his theory can be more favorably considered without this type of commitment to the ontology of "true" and "false", in my presentation of his views I shall abstract from them.

[3] It is not clear what Biegański means by claiming that truth is a judgment about another judgment. His objection to treating truth as an objective property of judgments does not exclude the possibility of treating truth as a relative property. As will be seen from what follows, even though Biegański does not acknowledge it, that's what he seems to be doing. Thanks are due to Sandra Lapointe for making me clarify this point.

the case, it must follow from the principles, and the judgments from which it follows must furthermore be true. Biegański's treatment of falsity is analogous:

> Falsity is a judgment asserted about another judgment, that the content of the latter is not justified, i.e. does not agree with the principles[4] (151)

This account of truth and falsity leaves room for something intermediate:

> [A] judgment stating about another judgment, that the content of the latter is not sufficiently justified, i.e. does not fully agree with [its] principles may be called probable (1910, 151).

Biegański does not speak about probability or doubt in the sense of the absence of a truth-value. Probability is a positive characteristic of the relation of a judgment to its principles. To make a judgment of probability or doubt about another judgment, we must be aware of the fact that the judgment in question is neither justified by its principles nor in conflict with them. So Biegański's logic of truth is a three-valued, or in fact many-valued logic since it allows for degrees of probability. It is a strong many valued-logic in Łukasiewicz's sense, and not in the sense of Kleene who treated his third value as the absence of value or in fact the absence of our knowledge about the value.[5]

To illustrate these concepts, Biegański uses the example of seeing somebody indistinctly from a distance (1910, 151). In the situation he describes, if I utter the judgment:

I think that my father is coming.

I express the probability of the judgment

(*) *My father is coming.*

I am aware that my impressions – what I see – do not support (*). But if the person comes closer so that I have more impressions upon which to base my judgment, I may have enough impressions to justify (*). This can be expressed by the judgment:

So it is true that my father is coming.

On the other hand, my impressions may turn out to be in conflict with those connected in my memory with my father. This can be expressed by the judgment:

So it is false that my father is coming.

According to Biegański, truth consists in the complete justification of a thought and such justification comes from impressions and other judgments whose

---

[4] Biegański in fact uses here the word "błąd", whose standard translation is "error". But because he uses it consistently in a way we would nowadays speak about truth and falsity instead of truth and error, I translate "błąd" as "falsity" in this context.

[5] On the concept of strongly three-valued logic and the difference between Łukasiewicz's and Kleene's logics see my (Kijania-Placek, 2002).

justification can in most cases be traced back to impressions. Since he considers even concepts to be accumulated experience, we may want to ask whether the difference between the idea that truth is based on correspondence with reality and the idea that truth is correspondence with knowledge of reality does not collapse, which seems to be a consequence of Biegański's position. In fact, Biegański explicitly claims that there is no contradiction between his concept of truth and the concept given by Aristotle, as long as reality is identified with the impressions and feelings we experience (1910, 162). But there is a difference, and in order to understand what this difference is we must consider the theory of cognition on which Biegański's conception of truth is based.

The main thesis of Biegański's theory of cognition consists in saying that we do not aim through the process of cognition at complete knowledge of reality but at the prediction of future events and properties of objects. Cognition is subject to evolutionary principles. Biegański's arguments are based on a clearly stated assumption "[that] the cognitive process must play a biological role in human life and that it has been appropriately adapted to it" (1910, 123). At least three arguments based on this assumption are worth mentioning. First, pointing to the way human senses actually work, their selectiveness and sensitivity to stable impulses, Biegański argues that we can see a concordance between their performance and the goal of cognition only if this goal is the prediction of future events. While less selective detection of impulses, or indifference in detecting stable and changing impulses would be an obstacle in prediction, they would be desirable if the goal were considered to be complete knowledge of reality (1910, 32, 51, 78). Secondly, considering the important role of abstract concepts in cognition – according to Biegański these concepts developed because they turned out to be indispensable in predicting individual properties of objects on the basis of previous observation of other individual properties – he argues that, since these concepts do not have counterparts in reality, explaining how they contribute to knowledge constitutes a major challenge to any theory that assumes that the goal of cognition is the reproduction of reality (1910, 44–45). He then argues that, when considering empirical judgments, we should acknowledge the fact that the judgments are not determined by the content of the impressions on which they are based, but by our predictive needs, which decide which correlations with former impressions we are aware of (1910, 22–23, 55–56). There may be more or less to the content of an empirical judgment than is justified by the actual impressions. Consider Biegański's example of the following judgment (1910, 155–156):

That object (over there) is a tree.

This judgment is based on visual impressions but, as such, is not completely justified since there are other impressions besides visual ones in the concept of a tree. I carry out this judgment because visual impressions are connected in my memory with other impressions that together form the concept of a tree. Following these associations, I can predict other features of the tree that might be important for my purposes. Biegański explains that most of our judgments based on impressions go beyond what we actually experience and claims that this extrapolation played a valuable role in our evolution, allowing for quicker orientation in our surroundings. Speed turned out to be more important than accuracy (1910, 56, 165–166). Yet, if the goal of cognition is not knowledge of reality but the prediction of future events and properties of objects, then reality should not be considered to be a criterion for assessing the value of cognition. Biegański explains the intimate bound between truth and cognition:

> Why is truth a positive value and falsity a negative one? It is because only true judgments lead us to correct predictions, while false ones almost always fail us. The truth and falsity of judgments, then, determine the value of prediction. [...] The goal of our cognition is true prediction (1910, 201–202).

And vice versa:

> [Truth] is only concerned with cognition; only knowledge expressed in judgments can be true or false. That is why we cannot introduce into the concept of truth elements that are beyond the limits of our knowledge. So, if by absolute truth we mean agreement with reality in itself, independent of our experience, then we introduce into the concept a fictitious element, which lies beyond the range of our knowledge, out of which our thought could not originate and by which it cannot be justified. So material absolute truth, in the sense given to it by logic, is impossible (1910, 163).

One should not here overplay the quasi-idealistic undertone. Biegański considered his concept of truth to be "objective" and even "absolute" in the sense defined by Twardowski in "On So-Called Relative Truths". According to Biegański, a thought is complete only if it is considered together with the principles, i.e. the thoughts and impressions from which it is inferred (1910, 211). Appealing to Twardowski's famous argument, Biegański claims that there is no contradiction between the thoughts expressed respectively by two sentence utterances in which two different speakers assert that the same glass of wine is at the same time sweet (for the first) and sour (for the second). These are different thoughts because their principles, i.e. the impression of sweetness for the first speaker and the impression of sourness for the second speaker, are different. So when its principles change, a thought changes, but the truth of the original thought remains unchanged. (1910, 177–179, 211) Likewise, two persons who base their judgments on the same principles cannot reach contradicting conclusions. So truth is also necessary (1910, 172, 179, 180). On the latter, however, Biegański's position differs from Twardowski's. The judgment expressed by the utterance:

It is raining

is not necessarily true for Twardowski, because it would be false if it was not raining. For Biegański, "It is raining" is a different judgment when it rains and when it does not rain since its principles and, in particular, the impressions on which it is based are in each case different. That identical principles are required for judgments to be identical unfortunately makes judgments trivially necessary: if true at all, all judgments are also necessarily true.

Biegański clearly sees that his conception of cognition is in opposition to the classical conception that treats complete and adequate knowledge of reality as the goal of cognition. Yet, he seeks to downplay the differences between his conception of truth and the classical conception by stressing the objective character of truth. Certain consequences of his conception of truth however indicate how important are differences. For one thing, Biegański believes that there are judgments that are neither true nor false. These, as we have seen, are judgments whose justification is not complete, which is the case for most empirical judgments, for instance. Besides, although truth as a complete justification is meant as an ideal for knowledge, the ideal is nonetheless attainable in particular instances (1910, 211). Let us consider a slightly simplified version of Biegański's favourite example:

There is an orange over there.

If I only see the object, then my judgment is not true but merely probable, because it is not completely justified. My concept of an orange contains features other than just those supported by my visual impressions, features such as its particular taste, origin, texture, etc. But on the other hand, the concept of an orange is also merely the result of accumulated experience and consists of a limited number of features. So it is in principle possible for me to have true judgments about oranges; I just need to check the object to obtain the impressions that would support the whole content of my concept of an orange, i.e. impressions that support the features that the concept of an orange consists of. By contrast, according to Biegański, under the classical conception of truth, knowledge is unattainable. Biegański believes that since the classical theory of truth holds that an object falls under the concept of an orange if and only if it is a real orange, then we're never justified in treating such judgments as true, because there is always a chance that we be fooled by somebody or, more importantly, that we be fooled by nature. It might be that there are fruits in all respect similar to oranges except for some feature that were never encountered before.[6]

What was the reception of Biegański's ideas in Poland? References to his work on truth among his contemporaries are rare and the only review I know of,[7] by Stanisław Leśniewski in 1913, is ironic and hostile. Leśniewski writes:

> The issue regarding the truth of certain claims and the correctness of their justification can only be settled once they are stated as clearly and unambiguously as possible [...]. This crucial condition of clarity and non ambiguity is, in my opinion, evidently not fulfilled by the reviewed work of Biegański: in both parts of his book we encounter over and over again expressions of which we cannot possibly know what they are concerned with. (Leśniewski 1913, 140)

Even though we may agree that many of Biegański's claims are controversial and that their formulations sometimes require clarification, his position is, contrary to what Leśniewski implies, relatively clear and philosophically interesting. This, however, cannot be said of Leśniewski's review who remained shallow, failed to give any examples supporting his objections and clearly violated the basic principle of charity.

---

[6] "Orange" is understood as a natural kind term for the sake of this argument.

[7] I am indebted to Jan Woleński and Arianna Betti for drawing this review to my attention.

13.3 Poznański and Wundheiler

I will now turn to two authors whose conception departs even more radically than Biegański's from the classical theory of truth, namely Edward Poznański and Aleksander Wundheiler. Both were members of the Lvov-Warsaw School and both were students of Tadeusz Kotarbiński. Their papers, for the most part, discuss various aspects of the methodology of science. Poznański was however interested in the philosophy of language as well.

In a paper entitled "The Notion of Truth in Physics" published in 1934 in a *Festschrift* for Kotarbiński, Poznański and Wundheiler develop a conception of truth which combines the coherence and the consensus criteria of truth and argue that there is no room for the notion of absolute truth in physics. They attribute truth to sentences (as opposed to judgements – or beliefs – and propositions), and by absolute truth they mean a combination of the following features: truth is independent of the knowing subject, it is independent from the truth of other sentences as well as from the state of science at a given time. Besides, absolute truth does not admit of degrees and is independent of the way in which it is verified (1934, 400). Poznański and Wundheiler claim that in physics and in other areas of human life where the classical notion of truth does not allow us to decide whether a sentence is true or false, we do not in fact resort to the notion of absolute truth, but to what they call the *operational notion of truth*. Under this conception, truth is directly connected with the methods by means of which we decide whether a sentence is true or false.

Poznański and Wundheiler's proposal combines two criteria: the coherence criterion, which, according to them, concerns a physical theory as a whole, and the consensus criterion, which concerns only the elementary sentences. With respect to the coherence criterion, they argue that physical theories form a system of interconnected sentences. Each sentence depends on others and there is no direction to this dependence. They contrast systems of that type with axiomatic systems in which there is only one direction of dependence: axioms form the basis on which all other sentences depend (1934, 406–407, 414). According to them, while axiomatic orders may be represented by a pyramidal structure, the system of physics should by contrast be schematized as a network of sentences. All of the sentences are interdependent. There are no axioms, in the sense of sentences that do not require justification. Even though we may admit of a distinction between experimental sentences and theoretical sentences, none of them can be treated as axioms, as each kind, according to Poznański and Wundheiler, need the other for their justification. This is why no sentence should be considered true or false in isolation. Truth applies to the system as a whole and consists in its coherence and comprehensiveness:

> [O]ur belief in the truth of particular sentences of the system is based on the fact that the system forms a coherent whole that contains all empirical facts known at a given time (1934, 417).

Poznański and Wundheiler are aware of the theoretical possibility of two or more systems which, while equally coherent and comprehensive, nonetheless contradict one another. They are ready to accept this consequence: the belief in one unique truth being, according to them, just a matter of faith (1934, 425).

Poznański and Wundheiler consider their conception to be not normative, but descriptive by nature. They claim to account for the actual use of the notion of truth in physics. In this respect, they are forced to accept a somewhat controversial consequence: since all sentences are interconnected, they can never be verified in isolation, and the process of verification presumably does not have a limit. But since an infinite process of verification is humanly impossible, this conception, as it stands, can hardly be considered an adequate *description* of the actual use of the notion of truth in physics: scientists seem to think that at least some of their claims are finitely verifiable. It is because of these considerations that Poznański and Wundheiler take the step of supplementing the coherence criterion of truth with the consensus criterion. According to them, the latter pertains to elementary sentences. Reference to elementary sentences does not here imply a contradiction with the above mentioned idea that no statements in physics are basic: Poznański and Wundheiler claim that the consensus criterion should, whenever a conflict arises, be subordinated to the coherence criterion. There are four types of elementary sentences according to them:

1. Sentences about reciprocal spatial relations – the occurrence or non-occurrence of spatial coincidences.
2. Sentences about reciprocal temporal relations – the occurrence or non-occurrence of temporal coincidences and the temporal order of phenomena.
3. Sentences about (physical) numbers, that is, about the equinumerability of two sets. [...]
4. Sentences about any other non-spatiotemporal coincidences, that is, about the absence of differences in other areas, e.g. identity of colour, lighting, pitch, etc. (1934, 430).

Poznański and Wundheiler, inasmuch as the consensus criterion of truth is concerned, follow fairly closely Norman R. Campbell, who, according to them, treated universal agreement as a criterion of truth. In fact, for Campbell, universal agreement is a criterion only for the choice of the subject matter of a particular science such as physics (Campbell 1920, 22). Campbell refrains from treating universal agreement as a criterion of truth (*ibid.*, 34). Hence, although Poznański and Wundheiler follow Campbell when it comes to determining, for instance, which sentences are to be accepted as elementary (the first three kinds of sentences come from Campbell; the last is proposed by them), the recourse to universal agreement as a criterion of truth in physics was, strictly speaking, their own idea.[8]

The second category of sentences with regard to which, according to Poznański and Wundheiler, there exists universal agreement are logical laws and rules of

[8] Later Quine used universal agreement to define observation sentences as those "on which all speakers of the language give the same verdict when given the same concurrent stimulation" (Quine, 1969, 87).

inference. As a result, the property of being apt to universal agreement is inheritable: we assign it not only to elementary sentences but also to sentences deduced from elementary sentences.

The verification of elementary sentences is brought about through the reports of those who have observed the phenomena in question. Because Poznański and Wundheiler want to exclude situations in which the disagreement of incompetent or physically impaired agents would render the criterion ineffective, thus anticipating obvious objections, they endorse what they call "universal agreement with reservations". They formulate laws of "universal" agreement which stipulate, among other things, that "universal agreement excludes the reports of those who are abnormal, not disinterested, incapable of understanding" and that there exists universal agreement concerning the laws of logic, rules of inference, and elementary sentences (1934, 433–434). So universal agreement with reservations is the universal agreement of typical, competent, and disinterested agents. Here again they clearly depart from Campbell, who insists that when testing for universal agreement we cannot take into consideration only persons who satisfy given conditions but, on the contrary, that it is necessary to take into account everyone, even "infants and animals [...] so far as their opinion is ascertainable" (Campbell, *Physics*, 23). Campbell's radical position is the result of his believing that elementary sentences are observational sentences and that, with respect to them, everybody is "normal". Campbell was evidently an optimist.

Poznański and Wundheiler take the above laws to be statements of a descriptive rather than of a normative character. However, according to them, the fact that universal agreement is subject to certain laws allows it to be used as a criterion of truth (1934, 434). Poznański and Wundheiler consider potential objections to accepting universal agreement as a criterion of truth. First, there might be cases where a sentence obtains universal agreement, although later, as science develops, its contrary comes to be accepted by the majority. If truth is a result of consensus, then the latter sentence would also be true. They answer by noting that there are virtually no instances of the overturning of unanimity once it has been achieved, especially if we bear in mind that the criterion applies to elementary sentences and not to scientific hypotheses, such as the shape of the Earth or the laws of physics. Second, one can imagine that we obtain universal agreement on false sentences. After all, we cannot exclude the possibility that "even with regards to elementary sentences 'everyone be mistaken' " (1934, 436). The answer to this objection is based upon their notion of operational truth, which the authors consider to hold in physics. They believe the notion of absolute truth to be "methodologically useless" and claim that its exclusion from dictionaries would change neither the content nor the value of the sentences of physics. Instead, they maintain that "the truth of sentences is defined directly by the methods of verification". They call the kind of truth defined by such processes *operational truth*, by analogy to the operational view of the meaning of physical concepts, according to which, for instance, "a physical magnitude is defined by the set of measurements which lead us to assign it a concrete numerical value" (1934, 440–441). According to this view, asserting that " 'a sentence is true' means operationally that 'a sentence is in agreement with the system to which it belongs', or that 'a sentence has obtained universal agreement' " (1934, 441). On this basis however it is not particularly clear what it might mean to say that such a sentence is false, if the only verification process available confirms the truth of this sentence. Furthermore, since the acceptance of a truth that is independent of the knowing subject (i.e. independent of the methods of verification available to the subject), makes it impossible to uncover errors, in practice this kind of truth remains a matter of faith and in the area of physics, the above objection is operationally meaningless.

The last objection, or perhaps reservation Poznański and Wundheiler consider is the question as to which criteria ought to be used in choosing or evaluating the laws of universal agreement. They opt for the coherence criterion of truth, which applies to the system of reality as a whole. According to them the criteria of truth are not to be found outside of the system, but belong to it and are to be evaluated through the system's coherence. Poznański and Wundheiler maintain that "we remain within our system, even when we speak about it" (1934, 438). Here they depart significantly from Campbell who excludes even the sentences of logic from the subject of investigation of the particular sciences.

One consequence of this operational view is that the truth of a sentence depends on the system to which it belongs, i.e., on the state of knowledge at a given time. As a result, it may change when the system changes. Also, truth becomes a matter of degree. While deduction transfers the certainty of elementary sentences to sentences deduced from them, in the process of induction which is used in building a physical theory, certainty, if there is such a thing, would be at best gradual. Moreover, as we have seen above, this view makes it possible for two conflicting truths to coexist, since it is possible for two systems, both explaining reality, to coexist. Poznański and Wundheiler accept these consequences and maintain that "[o]perational truth inherits only the indistinctness and fuzziness that are characteristic of the reality we come to know" (1934, 447). They also note that these "unfortunate" consequences concern foremost physical hypotheses and have little effect on the elementary sentences to which the consensus criterion of truth applies.

13.4 Conclusion

What are the common non classical features of the two conceptions presented above? The first and most obvious is the rejection of the notion truth as independent of the knowing subject and of the system of judgments to which it belongs. A less obviously but equally common feature is the role of consensus as a criterion of truth for elementary sentences and the laws of logic. Biegański explicitly rejects consensus as a criterion of truth, arguing that universal agreement is only a sign of the common, shared principles upon which the judgment is based. Those principles could always turn out to be false. If, however, we remember that according to Poznański and Wundheiler, consensus can be treated as a criterion of truth only for elementary sentences and the laws of logic, the disagreement vanishes. Elementary sentences are based on impressions and their content does not transcend what follows from them. The laws of logic, on the other hand, are justified *a priori*. So Biegański would agree that, inasmuch as the range of judgments to which Poznański and Wundheiler restrict the consensus criterion of truth is concerned, truth indeed follows from universal agreement.

What about the "laws of logic"? For Biegański they are justified *a priori*, while for Poznański and Wundheiler there is universal agreement regarding them. In this respect, both views agree on the truth-evaluative status of the laws of logic. Both also agree on the status of the law of bivalence, which is valid neither for Biegański nor for Poznański and Wundheiler. According to Biegański, there are judgments that are neither true nor false; probable judgments are an example. According to Poznański and Wundheiler, a hypothesis is neither true nor false if it is independent of the system of physics. And in this, both positions are in clear opposition to the view that was dominant in the Lvov-Warsaw School.[9]

Bibliography

Biegański, Władysław (1910), *Traktat o poznaniu i prawdzie* (Treatise on cognition and truth), Warsaw.

Biegański, Władysław (1908), *Logika medycyny, czyli krytyka poznania lekarskiego* (The logic of medicine, or a critique of medical cognition), Warsaw. A first edition was published as *Logika medycyny, czyli zasady ogólnej metodologii nauk lekarskich* (The logic of medicine, or the principles of the general methodology in the medical sciences) in 1894.

Biegański, Władysław (1915), *Teoria poznania ze stanowiska zasady celowości* (Epistemology from the point of view of the teleological principle), Warsaw, Towarzystwo Naukowe Warszawskie.

Campbell, Norman Robert (1920), *Physics. The Elements*, Cambridge, Cambridge University Press.

Kijania-Placek ,Katarzyna (2000), *Prawda i konsensus. Logiczne podstawy konsensualnego kryterium prawdy* (Truth and consensus. The logical basis of the consensus criterion of truth), Cracow, Jagiellonian University Press.

Kijania-Placek , Katarzyna (2002), "What difference does it make: three truth-values or two plus gaps?", *Erkenntnis* 56, 83–98.

Leśniewski, Stanisław (1913), 'Review of Władysław Biegański, *Treatise on cognition and truth* (in Polish)', *Wychowanie w domu i szkole. Czasopismo pedagogiczne* (Upbringing at home and at school. A pedagogical journal), Warsaw 1913, 139–142.

Poznański, Edward and Aleksander Wundheiler (1934), 'Pojęcie prawdy na terenie fizyki' (The notion of truth in physics), in E. Geblewicz et al. (eds.), *Fragmenty filozoficzne, I. Księga pamiątkowa ku uczczeniu piętnastolecia pracy nauczycielskiej w Uniwersytecie War-szawskim Profesora T. Kotarbińskiego* (Philosophical Fragments, I. Essays in honor of Professor T. Kotarbiński on the occasion of 15 years of his teaching at the Warsaw University), Warsaw, 97–143. Reprinted in T. Pawłowski (ed.). *Logiczna teoria nauki* (The logical theory of science), Warsaw. PWN, 1966, 399–448.

Quine, Willard Van Orman (1969), 'Epistemology naturalized', in *Ontological Relativity and Other Essays*, New York, Columbia University Press, 69–90.

[9] Łukasiewicz's rejection of bivalence is the only widely known example of departure from Twardowski's, Leśniewski's and later Tarski's views on the subject within the School.

Twardowski, Kazimierz (1900), 'O tzw. prawdach względnych' (On so-called relative truths), in *Księga pamiątkowa Uniwersytetu Lwowskiego ku uczczeniu pięćsetnej rocznicy fundacji Jagiellońskiej Uniwersytetu Krakowskiego* (Memorial book of the Lvov University in commemoration of the 500th anniversary of the Jagiellonian foundation of the University in Krakow), Lvov.

Woleński , Jan and Peter, Simons (1989), '*De Veritate*: Austro-Polish Contributions to the Theory of Truth from Brentano to Tarski', in K. Szaniawski (ed.), *The Vienna Circle and the Lvov-Warsaw School*, Dordrecht, Kluwer, 391–443.

Chapter 14
Leon Chwistek's Theory of Constructive Types

Bernard Linsky

From the readily available sources in English one can learn that Leon Chwistek was born in 1884 in Zakopane, studied logic at Göttingen briefly during 1908 and 1909, at Krakow under Ślezyński and Zaremba, and then taught in a secondary school in Krakow for several years.[1] After 1929 Chwistek was a Professor of Logic at the University of Lwów in a position for which Alfred Tarski had also applied. His interests in the 1930s were in a general system of philosophy of science, published in 1948 in English as *The Limits of Science*. Chwistek was also a painter in the Polish "Formist" school of expressionism and figure in the artistic scene of Poland between the world wars. He died in Moscow in 1944 having gone to Russia when the Germans invaded Poland in 1939. This broad outline of his career and work are all that would be known among English speaking philosophers who looked in the few familiar sources on Polish logic.[2]

For logicians, however, there is more material available. Chwistek's "scientific correspondence" with Bertrand Russell has been published, and one of his papers is included in Storrs McCall's collection *Polish Logic: 1920-1939*.[3] Chwistek's paper "The Theory of Constructive Types", published in two parts in 1924 and 1925, has scarcely been read at all, though it is widely cited. Most prominent of these citations is in the Introduction to the second edition of *Principia Mathematica* where Chwistek is mentioned for his proposal to reject the axiom of reducibility. Chwistek is now known among logicians primarily for his unsuccessful argument that the

B. Linsky (✉)
University of Alberta, Edmonton, AB Canada
e-mail: bernie.linsky@ualberta.ca

[1] I am grateful for many helpful comments on this paper from participants at the conference from which this volume derives, including Peter Simons, Gören Sundholm and especially Jan Wolenski.
[2] This biographical material is in fact gathered from Jordan in (McCall 1967) and (Grattan-Guinness 2000), (Feferman and Feferman 2004), and web sources. In discussions after the presentation of this paper, Jan Wolenski described Chwistek as a prominent member of mathematical and artistic circles in Krakow and later Lwów, and while a colorful figure, not the eccentric suggested by some references to his work.
[3] Jordan's historical article at the end of that volume says that Chwistek's early papers cleared up some notational problems in PM and invented the simple theory of types. Jordan is critical of Chwistek's later work.

S. Lapointe et al. (eds.), *The Golden Age of Polish Philosophy*, Logic, Epistemology, and the Unity of Science 16, DOI 10.1007/978-90-481-2401-5_15,

axiom of reducibility reintroduces semantic paradoxes, a criticism later revived by Irving Copi and answered by John Myhill and Alonzo Church in well known papers. In general, then, Chwistek is seen as a marginal to the great Polish Lwów-Warsaw school of logic; as Ivor Grattan-Guinness describes him in a chapter title, a "Pole apart" (Grattan-Guiness, 495–497). In his own study of the school, Jan Wolenski briefly mentions Chwistek, but passes over him as "in no way" a part of the Lwów-Warsaw group (Wolenski, 310).

Less well known are two papers, "Über die Antinomien der Prinzipien der Mathematik" from 1922, and "The Theory of Constructive Types" from 1924–1925. These two papers alone are enough to show that Leon Chwistek should have a place in any account of the story of Polish Logic in the "Golden Age." To begin with, these papers explain why Chwistek is seen as being out of the mainstream of Polish Logic. His work in the early 1920s was on the type theory of *Principia Mathematica* and the intensional logic to which it leads, which was very different from the interest in axiomatic set theory and mereology of the rest of the school. Chwistek's interest in syntactic issues and his inclination towards constructivism and even nominalism were in keeping with the main body of the Lwów-Warsaw school, but he differed from the majority in not adopting extensional logic. Chwistek's work is thus of more importance for studying Russell's work on logic during the twenties than that of the school working around him in Lwów.

14.1 The Letter of Recommendation for the Chair in Lwów

An examination of Chwistek's work in these papers provides an explanation for an otherwise puzzling incident that is a challenge for any assessment of Chwistek's reputation. This is the the competition in 1929 for the newly introduced chair in Logic at the university in Lwów in which Bertrand Russell wrote a letter supporting Chwistek over the other candidate, Alfred Tarski. On December 23rd of 1929 Bertrand Russell wrote to "Prof. Żylinski, Dean of Faculty of Mathematics – Lwów":

> I much regret that owing to my absence in America your letter of the 31st of October has remained hitherto unanswered. I know the work of Dr Chwistek and think very highly of it. The work of Mr. Tarski. I do not at the moment remember, nor have I access to it at present. In these circumstances, I can only say that in choosing Dr Chwistek you will be choosing a man who will do you credit, but I am not in a position to compare his merits with those of Mr. Tarski
>
> (Jadacki, 1986, 243).

The competition for the chair naturally led to strong feelings and partisanship among a number of prominent members of the philosophical and mathematical communities. For those unfamiliar with Chwistek and those personal relationships, this letter sounds peculiar. However, after investigating the nature of Chwistek's work on *Principia Mathematica*, and Russell's knowledge of Tarski's work at the time, this letter makes better sense. Now Russell must have known something of Tarski in 1929, for he had listed three papers published in *Fundamenta Mathematica* in the Introduction to the second edition of PM published in 1925.[4] Tarski's papers between 1924 and 1929, however, were on set theory and did not deal with logic. It is understandable, then, that he might have thought well of Tarski, but not been able to recall his work in logic when writing the letter away from his books. On the other hand Russell must have thought well enough of the works of Chwistek to still have Chwistek's work in mind. What exactly, then, was the "work of Dr Chwistek" of which Russell speaks? The two works cited in PM are the topic of this essay; "The Theory of Constructive Types" and "Über die Antinomien der Prinzipien der Mathematik."

14.2 Who Read "The Theory of Constructive Types"?

The introductory section "B" of Chwistek's "The Theory of Constructive Types" makes three significant points that have since become part of discussions of the technical aspects of PM. One was later described by Rudolf Carnap in *Meaning and Necessity*. The problem is that the no-classes theory in $*20$ of PM allows for ambiguities of the scope of class terms. The sentence '$\varphi\hat{x} \neq \hat{x}\varphi x$' has one reading on which it is true, as φ is not identical with some function coextensive with φ, yet it is false on the reading on which it means that some function coextensive with φ is not self-identical. Chwistek proposes an adequate solution in the form of explicit scope indicators, modeled on the scope indicators for definite descriptions which are presented in $*14$ of PM. The expression '$\phi(\hat{x}\psi x)$' which says that the class of ψs has the property ϕ should be replaced, he argues, by one which explicitly indicates that 'ϕ' is the scope of the class term: '$[\hat{x}\phi x].\phi(\hat{x}\psi x)$'. What is remarkable is that Chwistek's argument, and even the example he uses are very similar to points made by Carnap in *Meaning and Necessity*, and Quine in correspondence with Carnap in 1943 and even Gödel, briefly in his well known discussion "Russell's Mathematical Logic" in 1944. Yet these authors do not seem to have known of Chwistek's discussion.[5]

---

[4] These were: "Sur le terme primitif de la logistique" (1923), "Sur les 'truth-functions' au sense de MM. Russell et Whitehead" (1924), and "Sur quelques théorèmes qui équivalent à l' axiome du choix" (1924), all in *Fundamenta Mathematica*. The first two were combined in the English "On the Primitive Term of Logistic", in Tarski's papers *Logic, Semantics, Metamathematics*. Feferman and Feferman suggest that while the selection of Chwistek for the chair could be understood given Chwistek's seniority over the then quite young Tarski, and his work on the theory of types, there still might have been some element of anti-semitism in the selection. Hugo Steinhaus, who was Jewish, himself preferred his brother-in-law Chwistek over Tarski, so the issue couldn't have been seen as that by all at the time. See (Feferman and Feferman 2004, 66–68).

[5] References to the later history of this problem, and its appearances in Quine, Carnap and Gödel, can be found in my Linsky (2004a) and Linsky (2004b) In his well known paper on scope and definite descriptions, Arthur Smullyan mentions that there will also be issues about scope in the no-classes theory, citing Carnap, using the same notation for the scope of class terms modeled on that for definite descriptions that Chwistek proposes. My thanks to Leonard Linsky for remembering this detail of Smullyan's paper.

Who did know of Chwistek's argument? Certainly Russell must have appreciated these points? Notice, however, that the wrong name is given for the journal in the list at the end of the Introduction to the second edition of PM, namely "Annales de la Société Mathématique de Polonge".(Even though Chwistek called it "Annales de mathematique" in a letter to Russell of 22.11.1925.) The journal itself used the name "Annales de la Société Polonaise de Mathématique". It looks like neither Chwistek nor Russell had the published version before them when writing. It is likely that Russell read a large manuscript, one that Chwistek was trying to get published in various pieces in journals such as *Fundamenta Mathematicae*, and that he relied on Chwistek's report of where it would be published for the Introduction. It is almost certain, however, that the manuscript Russell saw would have included the discussion of the no-classes theory. Still, Russell never mentions any of those three points in the correspondence, or elsewhere.

An undated letter in the Russell Archives from Ramsey to Russell about the printers' proofs of the second edition of PM, (estimated to be after 20.2.24 and before 3.12.24) says: "I have read the introduction and verified nearly all the references." An accompanying sheet titled "Corrections" includes some proposed additions; papers by Weyl, Brouwer, and "Schönwinkel", with some corrections to the entries to Brower, "Tajtelbaum - Tarski" and Sheffer, which were presumably listed by Russell already, and a remark: "I have not checked Lewis, or Chwistek in the Annales de la Societé Mathématique de Pologne which is not in Cambridge." That particular issue might not yet have been published, or, more likely, no one in Cambridge subscribed.[6] So, even though Russell mentions Chwistek's project of abandoning the axiom of reducibility as "heroic" in the Introduction, there is no discussion of his various points about scope and propositional functions, and no extant evidence of his having studied it.[7]

Another opportunity for a citation comes in notes Carnap made in the 1930s on "Über die Antinomien...". The notes simply repeat Chwistek's description of "The Theory of Constructive Types" as not published, suggesting that he was not aware of its subsequent publication. Since both Gödel and Carnap, apparently independently, both later make Chwistek's point about scope, it is ironic that in correspondence between Carnap and Gödel, there is criticism of Chwistek's clarity in technical

---

[6] An exchange copy would have gone to Vienna, looking at the list of exchange subscriptions in the back of the issue. The names of Weyl, Brouwer, and "Schönwinkel" would seem to settle the question, for these cases, of which names were due to Russell and which added by Ramsey or others. See Feferman and Feferman 2004, 68 note 19. (Ramsey also suggests that Russell add Wittgenstein's *Tractatus* to the list, even after Russell had discussed it extensively in the body of his Introduction to the new edition.)

[7] Russell says: "Dr Leon Chwistek took the heroic course of dispensing with the axiom without adopting any substitute; from his work, it is clear that this course compels us to sacrifice a great deal of ordinary mathematics". (PM, xiv). In his review Church, remarks however, that "With the introduction of number theory by means of special axioms, his course is found to be less heroic than might be inferred from the comment of Russell, who, it would seem, had not seen Chwistek's paper in its final form."(170)

matters, but no mention of "The Theory of Constructive Types".[8] It seems then, that the most that can be asserted is that Chwistek anticipated this point, however close the similarity with later rediscoveries.

14.3 The Introduction to "The Theory of Constructive Types"

In addition to the problem about scope in the theory of classes, Chwistek makes two other very good points in the introductory section of the paper. One is a passing reference to scope problems with Whitehead and Russell's device for indicating propositional functions, specifically, their use the use of a caret over a variable as in '$f\hat{z}$'. This notation is also subject to scope ambiguities, and so Chistek proposes the notation such as '$\hat{z}f\hat{z}$' and '$\hat{x}\hat{y}R\hat{y}\hat{x}$', which is close to the later lambda notation: '$\lambda x f x$' and '$\lambda x \lambda y R y x$'.[9]

The third interesting feature of the introduction is how Chwistek shows an understanding of a point that has only recently been noticed in studies of the no-classes theory. Chwistek appreciates that Russell's notation for functions of classes '$f\hat{\alpha}$', following the use the Greek 'α' to range over classes, must nevertheless in fact be seen as functions of functions to propositions about classes defined by those functions, as now would be expressed by '$\lambda\phi f\{x : \phi x\}$'. These are very precise points about logical syntax, quite out of keeping with Chwistek's unfortunate reputation with Gödel and others for unclarity in matters of syntax.

Chwistek's paper is of particular interest for Russell scholarship because it shows that Russell was informed of rather basic, but correctable, errors in the system of the first edition of PM just as he was working on the second edition. Why did Russell ignore these points in Chwistek's paper? That these points were not mentioned by Russell shows something about Russell's attitude towards accepting the principle of extensionality, with the accompanying loss of the need for a

---

[8] In a letter to Carnap in April of 1932 we find "With this letter I am sending back to you Chwistek's offprint and manuscript with best thanks. Unfortunately I cannot write an elucidation about the latter, for most of what is in it (and precisely what is most important, e.g., type theory and real numbers) is treated in such a vague form that one can get nothing out if it without knowledge of the original works." (Gödel 2003, 345). Gödel is most certainly referring to later papers, but the suggestion is that Chwistek's style in technical matters is not precise.

[9] Chwistek's example is to consider a sentence produced by a predicative function g applied to a function f of two arguments, a variable of the type of $\phi!\hat{x}$ and a constant 'a', namely $g!\{f(\hat{\phi}!\hat{z}, a)\}$. Abstracting the '$a$', to get an expression for a function, we get an ambiguous expression $g!\{f(\hat{\phi}!\hat{z}, \hat{x})\}$ which could be either what was sought, a function which when applied to a gives $g!\{f(\hat{\phi}!\hat{z}, a)\}$, or else a sentence, which results from a function g applied to the result of applying f to two arguments, $\phi!\hat{z}$ and x. Of course if the logical type of the function g is specified, i.e., as either a function of propositions, or a function of monadic propositional functions, the ambiguity would be resolved. More recent examples of this scope problem, such as that from Hatcher (1968), avoid this resolution of the ambiguity by having f be a function of two variables, thus starting with a formula like $g[F(x, y)]$ and then demonstrating that the resulting $g[F(\hat{x}, \hat{y})]$, is ambiguous between $\lambda x g[\lambda y F(x, y)]$ and $\lambda y g[\lambda x F(x, y)]$. Distinguishing the logical types of g and F does not resolve this ambiguity. See (Hatcher 1968, 126).

no-classes theory of classes, corrected or not. The reason may be that Russell had already adopted the principle of extensionality by 1923. The possible distinction of truth value of different scopes disappears for both (proper) definite descriptions and classes if there are no intensional contexts. As well, in an extensional logic propositional functions are identical if coextensive, and so can simply be identified with classes. There is no need for the "no-classes theory". The lack of recognition of the other point, about the ambiguity of the abstraction notation, is not so easy to understand.

This is not a problem specific to an intensional logic. Any logic which allows higher order functions to apply to lower level functions will create the occasion for the ambiguity. It has been asserted, however, that Russell's logic, from the time of the *Philosophy of Logical Atomism* lectures, was committed to all atomic facts being first order. There might be higher level functions but these would be somehow derivative. Russell's example of analyzing " 'Before" is a relation' as 'If I say that x is before y, I assert a relation between x and y' certainly leads to this assumption.(Russell, *Logical Atomism*, 206). Perhaps pursuing the question of why Russell did not acknowledge the failure of the function notation may reveal something about his thinking about higher order functions in the second edition of *Principia Mathematica*.

Was the rest of Polish logic converted to extensionality by then? In a letter to Chwistek of 21.10.23 which prompted the reply quoted above, Russell says that he is already "incline[d] to the view (which I understand is also advocated by some in Poland) that all functions of propositions are truth-functions, and all functions of functions are extensional". Chwistek reports back that "The axiom of extension is here advocated by prof. Leśniewski and Mr. Tajtelbaum" (letter of 29.10.1923, Jadacki, 256.) He himself is not such an advocate, and that the consideration of an axiom of extensionality in his proposed system is a tentative addition to a basically intensional logic.

Consider finally a fourth valuable point of logical grammar from the introduction to Chwistek's 1924 paper. Logicians confronted with type theory have been concerned about the dual roles of predicate and general term, which seem to be played by the same entities. Frege confronted this as a paradox, that terms which seemingly name concepts, such as "the concept horse" are in fact unable to name concepts. More recently the idea that predicates appear as a subject of a first order sentence via their (quotation) names seems to lie behind the idea that Russell is somehow confused about the role of propositional functions as predicates or as entities, the source of the almost universal charge that his discussions of logic are marred by "sloppiness" about use and mention. Chwistek makes use of his new notation to make clear that the same expression can be both a general term and a predicate, although in logically different dress in each role. Thus his two definitional numbers (in his revised PM):

$$0 \cdot 16 \quad F(\lambda') \text{ iff } \hat{\lambda}[F(\hat{\lambda})]\{\lambda'\}, \text{ etc.}$$

$$0 \cdot 18 \quad \lambda \text{ stands for } \hat{\xi}[\lambda\{\hat{\xi}\}]$$

The former looks like lambda conversion, the latter like an explanation of the relation between predicates and second order terms in second order logic.[10] One can have a second order term (general term) which is to mean the same as the predicate from which it is derived.[11]

14.4 Extensionality and the "Pure" Theory of Types

Chwistek's plan for the technical body of his "Theory of Constructive Types" is to "...take directly from Principia all that remains true, if the axiom of reducibility is false and if functions of a given type are used as variables instead of matrices"(19). This he calls "*the pure theory of types*, or the *theory of constructive types*"(13). The reason for the term 'constructive' is Chwistek's view that the axiom of reducibility asserts the existence of a predicative function equivalent to a given function of higher type yet doesn't guarantee that the predicative function is definable. The axiom only asserts the existence of such a function without producing it, and so leads to a system which is not "constructive" in Chwistek's intended sense. Chwistek would have been aware of the notions of constructivism in both Poincaré and Weyl and does not propose to be introducing a new notion.

The notions of class and identity are heavily involved with the axiom of reducibility in PM. There identity is defined by:

$$*13 \cdot 01.\, x = y. =: (\varphi) : \varphi!x. \supset .\varphi!y \text{ Df}$$

This defines the identity of x and y as the sharing of all the same predicative functions φ. Given the axiom of reducibility, this has the force of the full identity of indiscernibles, i.e., x and y are identical if they share *all* properties. That cannot be stated as such in type theory as it would quantify over properties of all types.

Classes are defined with the famous "no-classes" theory, which uses a contextual definition of contexts in which class expressions occur in terms of properties of coextensive (and predicative) functions, much as the theory of definite descriptions allows for the "elimination" of descriptions from contexts in which they occur:

$$*20{\cdot}01 \quad f\{\hat{z}(\psi z)\}. = : (\exists\phi) : \phi!x. \equiv_x .\psi x : f\{\phi!\hat{z}\} \quad Df$$

Chwistek himself, however, makes "no difference at all between a function with I variable and a class, or between a function with II variables and a relation" (40). Thus a class $\omega_{(\alpha)}$ (of entities of type α) is simply indentified with the function ω:

$$20 \cdot 01\ \omega_{(\alpha)} =_{df} \hat{u}\,[\omega_\alpha\{\hat{u}\}]$$

---

[10] Chwistek credits Leśniewski for showing the need for such a principle as $0 \cdot 16$ ("Theory", 25).

[11] This paper is in Alonzo Church's "Bibliography of Symbolic Logic" (Church, 173) but Church makes no reference to anyone when later introducing the λ calculus.

This is involved with Chwistek's proposal to define identity as coextensiveness. Chwistek objects to what he terms the "Leibnizian" definition of identity in PM. Instead he proposes this definition of identity, *for classes*:

$$13 \cdot 01.\ \alpha = \beta.\ =_{df}\ \{\bar{x}\}.\alpha\{\bar{x}\} \equiv \beta\{\bar{x}\}.$$

Given Chwistek's identification of classes with functions, what he has done is to move from a definition of identity as belonging to all the same classes, to define identity as having all the same members.[12] The former, "Leibnizian", notion of identity (for entities of type a) Chwistek defines as:

$$13 \cdot 001 \quad (x =_L y)_a =_{df} (\bar{u}) : \bar{u}\{x\} \equiv \bar{u}\{y\} : \mathcal{T}\{\bar{u}, a\}.$$

('$\mathcal{T}\{\bar{u}, a\}$' indicates that the variable 'u' is of type a.) Chwistek then defines the notion of an *extensional* function:

$$13 \cdot 04 \text{ extens } [x_{(a)}] =_{df} \hat{f}[(\bar{u}, \bar{v})\ :\ \bar{u} \underset{\overline{a,u}}{=} \bar{v} \supset : \hat{f}_a\{\bar{u}\} \equiv \hat{f}_a\{\bar{v}\} : \mathcal{T}\{\hat{f}, x_{(a)}\} :]$$

This says that x (which is a function of type (a)), is extensional just in case it is of the same type as a function f which does not discriminate between coextensive arguments.

He then proves:

$$13 \cdot 12 \vdash : u = v\ .\supset :\ f_a\{u\}\ .\ \text{extens}[x_{(a)}]\{f\}. \equiv .\ f_a\{v\}\ .\ \text{extens}[x_{(a)}]\{f\}$$

In other words, extensional functions do not discriminate between identical classes (with identity so defined).

The "axiom of extension" is stated as

$$.(\bar{x}).\alpha\{\bar{x}\} \equiv \beta\{\bar{x}\}. \supset .\ f\{\alpha\} \equiv f\{\beta\} :$$

Chwistek sees the axiom of extension as embodying what he calls the "Realistic Hypothesis" about functions, which amounts to the possibility of having the same function denoted by two different expressions. The "Nominalistic system of Metaphysics (Ontology)" requires the impossibility of two class expressions denoting the same class, and hence the impossibility of proving that two equivalent classes are identical. Chwistek does not argue against the axiom of extension directly, saying only that:

[12] Robinsohn proves the independence of the two (stated as axioms), in Zermelo-Fraenkel set theory. In that paper Robinsohn (later 'Robinson') describes Zermelo's 1908 treatment of identity saying "Zermelo introduces equality intensionally – if two symbols x and y represent the same object, we write $= (x, y)$". Thus even in 1939 the Fregean tradition of treating identity as a logical notion independent of the theory of sets is considered "intensional" and to be improved upon. Chwistek doesn't consider the identity of individuals in his system.

> This axiom seems to have had great success in recent years. I never should care to discuss its truth. I am convinced that we never get a contradiction from using this axiom, but I am also convinced that its negation is consistent with the primitive propositions of the Logical Calculus and with the directions of the Pure Theory of Types.
>
> (Chwistek "Theory, II" 92–93)

All of this means that while Chwistek identifies classes and propositional functions, he introduces a defined notion of identity, which amounts to coextensiveness, in such a way that he can also identify extensional functions within his system. What Russell would have to prove about classes, using his no-class theory and the axiom of reducibility, Chwistek is able to prove about the extensional functions within his type theory without the axiom of reducibility. Russell is able to define '$x \in \varphi\hat{x}$' and still prove the "axiom" of extensionality as a theorem: classes are the same if they have the same members. Chwistek needs to restrict the membership relation $\in$ to extensional functions, and this he does in his $20 \cdot 02$.

In the second edition of *Principia Mathematica*, Russell proposes abandoning the axiom of reducibility while adopting the principle of extensionality. Russell says simply:

> Consequently there is no longer any reason to distinguish between functions and classes, for we have, in virtue of the above,
>
> $$\varphi x \equiv_x \psi x \,.\supset.\, \varphi\hat{x} = \psi\hat{x}$$
>
> We shall continue to use the notation $\hat{x}(\varphi x)$, which is often more convenient than $\varphi\hat{x}$; but there will no longer be any difference between the meanings of the two symbols. Thus classes, as distinct from functions, lose even that shadowy being which they retain in $*20$. (PM, xxxix)

Yet there is no indication that Russell proposes any modification of $*13 \cdot 01$ as a definition of identity, and so it is not clear just what force the identity assertion has. Perhaps Russell was relying on Chwistek's discussion here, which he mentions later in the "Introduction to the second edition." Russell may have realized that adding a principle of extensionality to the system of ramified intensional functions in the first edition was not so straightforward as he suggests.

Not only does Chwistek explicitly define the notion of an extensional function, and use it to produce his own "no-classes" theory, he also considers an additional "Axiom of Intensionality".

$$13 \cdot 002 \quad \text{Intax} =_{df} .(\hat{x}[\beta\{\hat{x}\}|\gamma\{\hat{x}\}] =_L \hat{x}[\beta'\{\hat{x}\}|\gamma\{\hat{x}\}]) \supset (\beta =_L \beta').$$

As the symbol '|' is the Sheffer stroke, primitive in the system, and so this is an inductive clause for a more general result involving any proposition in which β (and β') occur. In other words, then, the axiom of Intensionality amounts to the claim that (Leibniz) distinct functions lead to distinct compound functions.

Among applications of this axiom of intensionality is a proof of the axiom of infinity, one in a long line of attempts to show how in an intensional system one can prove that there are an infinite number of propositions or other intensional individuals.[13]

14.5 The "Simplified" Theory of Types

In his introductory "Critical Examination" of the system of Whitehead and Russell, Chwistek describes a "simplified" theory of types:

> Suppose we can speak about "all properties of x", i.e. about "all propositional functions $\Phi\hat{x}$ such that either Φx or $\sim \Phi x$". We shall have to deal 1^0 with individuals i.e. objects being neither propositions nor propositional functions; 2^0, with propositional functions whose arguments take individuals as possible values, i.e. propositional functions of the 1st type; 3^0, with propositional functions whose arguments take functions of the 1st type as possible values, i.e. propositional functions of the 2^d type ... and so one. Such a simple hierarchy of types would be, as a matter of fact, sufficient to build up a self-consistent system of Logic, there being no purely Logical paradoxes based on the idea of "all properties of x". Nevertheless, as this last idea does not exclude such contradictions, as Richard's paradox, or König's, it seems to be interesting to get a system of Symbolic Logic, free from such contradictions. To avoid these we must agree with Whitehead and Russell that the idea of "all properties of x" is meaningless.
>
> (Chwistek, "Constructive Types" 12)

Chwistek has the idea of the Simple Theory of Types, but sees that it is not applicable to the intensional and semantic paradoxes. In his review of Chwistek's work, from 1937, Alonzo Church credits Chwistek with having the idea of the simple theory of types on the basis of these sentences in "Antinomies":

> To eliminate this antinomy [of self-predication] it is sufficient to accept the simple theory of types which distinguishes individuals, functions of individuals, functions of these functions, and so on. The distinction of orders of functions of a given argument, thereby the introduction of predicative functions, and the consequent resort to the axiom of reducibility, is from this point of view a superfluous complication of the system. It should be noted that the elimination of these elements from the theory of types of Whitehead and Russell would render this theory exceedingly simple and manifest. If Whitehead and Russell decided not to carry out this simplification, they undoubtedly did so because of the conviction that a system of logic admitting Richard's antinomy could not be regarded as the final word on what it is possible to attain in this field.
>
> (Chwistek "Antinomies", 342–343).

Church presents the orginal, Polish, version of this paragraph in his review of "The Theory..." in the *Bibliography of Symbolic Logic* (Item 220).[14]

---

[13] See C.A. Anderson (1989) for a recent version of this. His system is "Russellian Intensional Logic".

[14] Church repeats this credit in his paper for the Congress of Unified Science in 1939, eventually published in 1976. After describing the ramified theory of types he says "The simple theory of types was proposed as a modification of this theory by Chwistek (1921) and Ramsey (1926), and was adopted by Carnap (1929), Gödel, Tarski, Quine and others. "Church argues that while Schröder clearly anticipated the simple theory in certain respects, Frege's notion of *Stufe* of concepts and concepts of concepts should *not* be seen as a theory of types of sets and sets of sets, etc.

Ramsey's well known distinction between semantic and logical paradoxes, and the subsequent defense of the simple theory of types to handle the paradoxes of set theory, is presented in his 1925 paper "The Foundations of Mathematics", which also includes a discussion of Chwistek's objections from 1922 to the axiom of reducibility. So Ramsey cites and discusses a paper in which Chwistek at least raises the issue of the two sorts of paradoxes. The 1922 paper, however, doesn't mention the simple theory of types, except indirectly, via its discussion of arguments against ramifying the theory of types. It looks as though Ramsey narrowly missed encountering Chwistek's idea of a simple theory of types. So much for the priority issue. There is nevertheless no doubt that Ramsey is the first to propose such a theory as an adequate type theory for logic, and to so explicitly defend it as part of a general solution to all of the paradoxes.

To be clear, the "constructive theory of types" is the PM system without the axiom of reducibility. The "Simplified theory", on the other hand, is the "simple" theory of types. This is the sense in which the simple theory does appear in Chwistek's paper. It is *not* the type structure for the "constructive" theory. Indeed, in "The theory of constructive types" Chwistek is more concerned with *objections* to the simplified theory:

> Note that the simplified theory of types, as expounded on p.12 of Part I, may be used in Mathematics without any risk of getting a contradiction. To avoid such paradoxes as Richard's or König's, it is quite sufficient to assume a direction excluding from the scope of the system any function which is not constructed with the symbols of the system itself. An analogous method is used by mathematicians dealing with the system of axioms of Zermelo. Such a method, though very convenient, is nevertheless inconsistent with certain fundamental problems of Logic and Semeiotics. Moreover the simplified theory of types implies the existence of functions which cannot be built up, unless we assume that all functions are extensional functions (the Axiom of Extension).
>
> ("Theory", II, 92–93)

The reference to Zermelo raises an issue of the day about set theory, which had clearly emerged as the rival to PM as a foundation for mathematics. What we now call the axiom (schema) of separation was given a second order formulation, saying that any "definite" property would determine a subset of an already established set. The notion of "definite" property was meant to exclude those that might introduce paradoxes by using notions of definabilty, truth, etc., which appear in the semantic and intensional paradoxes. There were several proposals about how to make the notion of "definite" property more precise, including Skolem's ultimately victorious proposal that definiteness be explicated as picking out those properties that can be expressed by predicates in the first order logic with identity, using 'ϵ' as the sole non-logical constant. There were other proposals, including that which Chwistek discusses here, which would amount to those properties that can be expressed with the notions of set theory.[15] Chwistek here is raising the point that the intensional and semantic paradoxes, which were seen as motivating the ramified theory of types,

---

[15] In particular Weyl who both proposes a "constructive" logic and gives a proposal for how to identify "determinate" properties on the basis of the linguistic resources used to express them.

could be avoided in the same way that they were in set theory of the time, by introducing a notion similar to that of "definite property". Still, that result is meant to be an objection to using simple type theory.

A second objection, which Chwistek mentions here, is that the simple theory of types without an axiom of extensionality would guarantee the existence of indefinable functions, and thus be in violation of the spirit which leads to the "constructive" theory of types. The argument is obscure. It goes as follows. Let:

$$G =_{df} \hat{\alpha}\hat{x}[(\exists \bar{f}).(\hat{z}[(\exists \bar{\beta})\bar{f}\{\bar{\beta}, \hat{z}\}] =_L \hat{\alpha}). \sim \bar{f}\{\hat{\alpha}, \hat{x}\}]$$

Suppose that

$$\sim G\{\hat{z}[(\exists \bar{\beta})G\{\bar{\beta}, \hat{z}\}], x\}$$

i.e.,

$$(\bar{f}).(\hat{z}[(\exists \bar{\beta})\bar{f}\{\bar{\beta}, \hat{z}\}] =_L \hat{z}[(\exists \bar{\beta})G\{\bar{\beta}, \hat{z}\}]) \supset \bar{f}\{\hat{z}[(\exists \bar{\beta})G\{\bar{\beta}, \hat{z}\}], x\}$$

by instantiating G for $\bar{f}$ we get:

$$\hat{z}[(\exists \bar{\beta})G\{\bar{\beta}, \hat{z}\}] = \hat{z}[(\exists \bar{\beta})G\{\bar{\beta}, \hat{z}\}] \supset G\{\hat{z}[(\exists \bar{\beta})G\{\bar{\beta}, \hat{z}\}], x\}$$

The antecedent is a truth of logic, so:

$$G\{\hat{z}[(\exists \bar{\beta})G\{\bar{\beta}, \hat{z}\}], x\}$$

that is (A),

$$[(\exists \bar{f}).(\hat{z}[(\exists \bar{\beta})\bar{f}\{\bar{\beta}, \hat{z}\}] =_L \hat{z}[(\exists \bar{\beta})G\{\bar{\beta}, \hat{z}\}]). \sim \bar{f}\{\hat{z}[(\exists \bar{\beta})G\{\bar{\beta}, \hat{z}\}], x\}$$

Chwistek concludes

> Now, it is obvious that the function f, whose existence is proved by (A), cannot be equivalent to G. Therefore we shall never have such a function, unless we assume the axiom of extension. As a system containing such an axiom is no longer one of pure logic, we see that there is no system of pure logic to be based on the simplified theory of types.
> (Chwistek "Theory", II, 96)

This reasoning is hard to follow. Presumably by "equivalent" above, Chwistek means "Leibniz" equivalent. Thus if we have some term instantiated for $\bar{f}$, say H, and $H =_L G$, then H can be substituted for G (because of their Leibniz identity), and so the conclusion leads to a contradiction. It is not clear what it means to say that the axiom of extensionality would guarantee that we "have" such a function, nor why the argument would not lead to a contradiction without the axiom of extensionality. It is thus uncertain what conclusion Chwistek should draw from this argument. Chwistek's insights may have been overlooked because they were buried among obscure arguments like this.

14.6 The Argument Against the Axiom of Reducibility

The history of Chwistek's criticisms of the axiom of reducibility is much better known. The thought is that while some higher order function such as "is a property asserted by such and such a predicate" will not apply to that very function, there will be, according to the axiom of reducibility, a lowest level, or predicative function coextensive with it to which it will apply. The door seems to open for the Richard and Grelling paradoxes of "the least number not nameable in fewer than 16 words" and the the predicate that does not apply to its own name.

Chwistek presents his argument in both of the papers from 1921 and 1922. The 1922 version "Über die Antinomen der Prinzipien der Mathematik" in *Mathematische Zeitschrift* is the one later discussed by Copi (1950–51) and Myhill (1974). (The 1921 paper in Polish, is translated as "Antinomies of Formal Logic" in McCall.) The "Antinomen" version is formalized showing that Chwistek saw this as a technical difficulty for the theory, and thus eventually leading to a technical solution by Church that advanced the study of the solution to the paradoxes in type theory.

Chwistek presents the following formula as defining a function f of an argument x (Chwistek, "Antinomen", 239):

$$x \in D`R. (\exists\alpha). xR\beta \equiv_\beta \beta = \alpha . \sim x \in \alpha$$

That is, x is in the domain of a functional relation R, which maps x onto a unique α to which x does not belong.[16] We consider the set of all such f (i.e. for the various instances of R) which are definable in some finite number of symbols (or syllables). Since there are only countably many definable expressions, we can consider the many-one relation S between numbers and and definable functions of the kind of f. When we let S be substituted for R above, we define the function $\Phi\hat{x}$. We have just defined $\Phi\hat{x}$, so S maps some number n onto it. By a familiar argument, it is easy to show that $\Phi n \equiv\sim \Phi n$, hence a contradiction.(The popularization of this uses 'x words are the least needed to name α' as f. This S is nameable, and by counting we determine that n is, say, 9, in this case, and so a paradox.) Now the ramified theory of types blocks this paradox, as it was indeed designed to do, by introducing *orders* for the various functions α and Φ to show how the latter predicate which is defined in terms of the bound variable 'α' will have to be of a higher order.

Chwistek's argument against the axiom of reducibility begins at this point. The axiom of reducibility guarantees that we can find a Φ which is of the same type as f, and so the paradox is regained.

Chwistek himself saw this as a *reductio ad absurdum* of the existence claim involved in the axiom of reducibility, an existence not backed up by an actual construction. In a letter to Russell, dated 29.10.1923, Chwistek says that the derivation of Richard's paradox '... seems however *to be no more actual*', presumably because

---

[16] Copi (1950–51) puts this in terms of a function ϕ which is true of x rather than a class α to which x belongs.

the derivation requires additional existence assumptions which are refuted by the derivation of an absurdity (Jadacki, 256).

The accepted response to Chwistek's argument is forshadowed in this remark of Russell in 1908, in his paper "Mathematical Logic as based on the Theory of Types":

> The essential point is that such results are obtained in all cases where only the truth or falsehood of values of the functions concerned are relevant, as is invariably the case in mathematics. Thus mathematical induction, for example, need now only be stated for all predicative functions of numbers; it then follows from the axiom of classes [the axiom of reducibility] that it holds of *any* function of whatever order. It might be thought that the paradoxes for the sake of which we invented the hierarchy of types would now reappear. But this is not the case, because, in such paradoxes, either something beyond the truth or falsehood of values of functions is relevant, or expressions occur which are unmeaning even after the introduction of the axiom of reducibility. For example, such as statement as 'Epimenides asserts ψx' is not equivalent to 'Epimenides asserts $\phi!x$', even though ψx and $\phi!x$' are equivalent.
>
> (Russell, "Mathematical Logic", 82–83)

While the axiom of reducibility guarantees that there is a predicative function equivalent to any that is presented, (S in this case) there is no problem with the low order of that function. We know that there is such a function, but do not have a definition of it, and so no formula that expresses it. In general, then, the axiom of reducibility will guarantee the existence of functions of the lowest order that are true of just the true formulas of a given type, or those definable, etc., but no semantic paradoxes are forthcoming because no formula is presented which expresses that predicative function.

Frank Ramsey explicitly discusses Chwistek's argument in "Foundations of Mathematics", seemingly following Russell's earlier line.

> Dr. L. Chwistek appears to have overlooked this point that, if a function is definable, the equivalent elementary function need not also be definable in terms of given symbols. In his paper "Über die Antinomen der Prinzipien der Mathematik" in *Math. Zeitschrift* 14, (1922), pp. 236–243, he denotes by S a many-one relation between the natural numbers and the classes defined by functions definable in terms of certain symbols. $\phi\hat{z}$ being a non-elementary function of this kind, he concludes that there must be an n such that $nS\hat{z}(\phi z)$. That is, however, a fallacy, since $nS\hat{z}(\phi z)$ means by definition
>
> $$(\exists\psi) : \psi!x \equiv_x \phi x \;.\; nS(\psi\hat{z})$$
>
> and since $\psi!\hat{z}$ is not necessarily definable in terms of the given symbols, there is no reason for there being any such n.
>
> (Ramsey, 28 note 2)

Ramsey thus takes the relation S to hold between numbers and classes, and so via the no-classes theory with its reduction of a relation, to a class defined by some arbitrary function to a relation to some predicative function coextensive with the original. So, Ramsey argues, the argument fails on the grounds that while S might be a definable relation to classes, it is not a definable relation to functions. However, S can be taken to be a relation to propositional functions. The precise statement of Ramsey's objection does not seem correct, even if it is in the right direction.[17]

Chwistek's argument eventually led to real progress in understanding the theory of types. This development came through a resurrection of the argument in a new form by Irving M. Copi. Copi gives full credit to Chwistek, and indeed includes a substantial discussion of Chwistek's argument. He takes Ramsey's view of Chwistek's original argument, that the coextensive property that exists is not necesarily definable, and since the function in Richard's paradox has to be definable, this refutes his argument. But Copi thinks that the Grelling (Heterological) version survives as all that is needed is that there be a function true of the same things as "heterological", and it is no part of the argument that 'heterological' is definable. He symbolizes the problematic sentences as $Het(`Het')$ and $\sim Het(`Het')$. The predicate 'Heterological' applies to those predicates which are not true of their own names. A simple example is that the predicate name 'Short' is indeed short and so the predicate 'Short' is not heterological. The predicate name 'Long', however is not long (indeed it is shorter than 'Short'), and so 'Long' is heterological. Now, is 'Heterological' itself Heterological, or not? A contradiction results. The predicate which is predicative and coextensive with 'Heterological' is describable, or nameable, in some sense, even if it is not explicitly defined or constructed. Copi's argument is that in order to avoid this paradox all possible semantic notions must be eliminated from the theory of types, thus making the axiom of reducibility "redundant", as it was introduced to eliminate the semantic paradoxes. If they can't even be formulated in the language of the theory of types, the introduction of the axiom is redundant.

The progress which resulted from this revival of Chwistek's argument came in the form of two well known papers, Alonzo Church's "Comparison of Russell's Resolution of the Semantical Antinomies with That of Tarski" in 1976 and John Myhill's "Refutation of an unjustified attack on the Axiom of Reducibility" in 1979. Church takes the challenge of Copi's article and shows just how much of semantics can be expressed in the ramified theory of types. In fact, it turns out, there are close parallels between Tarski's notion of a hierarchy of languages and metalanguages, and the definitions of semantic notions such as truth in the ramfied theory of types. There is a limitative result, however, which is that the "denotation", or naming relation 'Δ' which can be introduced into the theory of types, can be proved not to *represent* the naming relation in the technical sense that 'Het' is not provably named by any expression. The original reply to Chwistek's objection to the axiom of reducibility now reappears. This despite the fact that "Truth" (for a particular language) *is* definable in a suitably augmented ramified theory of types.

[17] At the conference where this paper was presented, Jan Wolenski questioned the cogency of Ramsey's argument. He pointed out that if one takes Chwistek to be arguing against the introduction of the axiom of reducibility to a ramified system of types of functions without the accompanying no-classes theory of classes from PM, Ramsey's objection would obviously miss the point.

14.7 Conclusion

The is an overview of the nature of Leon Chwistek's views of *Principia Mathematica*, some of the history of his interactions with Russell, and the later reception of Chwistek's views. No great revision of our views of the history of Polish logic is called for as a result of seeing these facts reported together, but our appreciation of the depth of activity in the field, even by a "Pole apart", should be enhanced. Chwistek can be placed in a stream of thought leading to the late work on intensional logic by Myhill and Church. He saw the possibility of a simple theory of types, but rejected it because it could not handle the intensional paradoxes. He also found scope problems with classes and functions which are due to intensional contexts, and investigated how to add an extensional theory of classes to type theory. Although such logical notions were out of the main stream of Polish Logic in the Lwów-Warsaw school, Leon Chwistek was nevertheless a real part of the "Golden Age of Polish Philosophy".

Bibliography

Anderson, C. Anthony. "Russellian Intensional Logic", in *Themes from Kaplan*, J. Almog, J. Perry and H. Wettstein, eds., New York: Oxford University Press, 1989: 67–103.

Carnap, Rudolf. Notes on Chwistek, Manuscript RC 081-25-01, Carnap Papers, Archive of Scientific Philosophy, University of Pittsburgh, 1932.

Carnap, Rudolf. *Meaning and Necessity*, Chicago: University of Chicago Press. 1947.

Church, Alonzo. "A bibliography of Symbolic Logic." *The Journal of Symbolic Logic*, 1, 1936: 121–192.

Church, Alonzo. Review of Leon Chwistek, *Überwindung des Begriffsrealismus*, in *Journal of Symbolic Logic*, 2, 1937: 168–170.

Church, Alonzo. "Comparison of Russell's Resolution of the Semantical Antinomies with that of Tarski", *Journal of Symbolic Logic*, 41, 1976: 747–760.

Church, Alonzo. "Schröder's Anticipation of the Simple Theory of Types", *Erkenntnis*, 10, 1976: 407-411.

Chwistek, Leon. "Antynomje Logiki Formalnej", *Przeglad Filozoficzny* 24, 1921: 164–71. Translated as "Antinomies of Formal Logic" by Z. Jordan in McCall, 1967, 338–345.

Chwistek, Leon. "Über die Antinomien der Prinzipien der Mathematik", *Mathematische Zeitschrift* 14, 1922: 236–243.

Chwistek, Leon. "The Theory of Constructive Types, (Principles of Logic and Mathematics), Part I", *Annales de la Société Polonaise de Mathématique (Rocznik Polskiego Towarzystwa Matematycznego)*, Vol. II (for 1923), 1924: 9–48.

Chwistek, Leon. "The Theory of Constructive Types, (Principles of Logic and Mathematics) Part II", Vol. III (for 1924), 1925: 92–141.

Chwistek, Leon. *The Limits of Science*, trans. E. Brodie, London: Kegan Paul, 1948.

Copi, Irving M. "The Inconsistency or Redundancy of *Principia Mathematica*.", *Philosophy and Phenomenological Research*, XI, 1950–1951: 190–199.

Feferman, Anita Burdman and Feferman, Solomon. *Alfred Tarski: Life and Logic*, Cambridge: Cambridge University Press, 2004.

Gödel. *Kurt Gödel: Collected Works*, Vol.IV, Solomon Feferman, et. al., eds., Oxford: Oxford University Press, 2003.

Grattan-Guinness, Ivor. *The Search for Mathematical Roots: 1870-1940*, Princeton: Princeton University Press, 2000.

Hatcher, William S. *Foundations of Mathematics*, Philadelphia: W.B. Saunders, 1968.
Jadacki, Jan. J. "Leon Chwistek - Bertrand Russell's Scientific Correspondence." *Dialectics and Humanism: The Polish Philosophical Quarterly*, XIII(1), 1986: 239–263.
Linsky, Bernard. "Leon Chwistek on the No-Classes Theory of *Principia Mathematica*." *History and Philosophy of Logic*, 25, 2004a: 53–71.
Linsky, Bernard. "Classes of Classes and Classes of Functions in *Principia Mathematica*", Godehard Link, ed. *One Hundred Years of Russell's Paradox*, Berlin and New York: Walter de Gruyter, 2004b: 435–447.
McCall, Storrs, ed. *Polish Logic: 1920–1939*, Oxford: Clarendon Press, 1967.
Myhill, John. "The Undefinability of the Set of Natural Numbers in the Ramified Principia" , in George Nakhnikian, ed., *Bertrand Russell's Philosophy*, London: Duckworth, 1974: 19–27.
Myhill, John. "A Refutation of an Unjustified Attack on the Axiom of Reducibility", in *Bertrand Russell Memorial Volume*, H.D. Lewis, ed., George Allen and Unwin, 1979: 81–90.
Ramsey, Frank P. "The Foundations of Mathematics", in *The Foundations of Mathematics and Other Logical Essays*, London: Routledge and Kegan Paul, 1931.
Robinsohn, Abraham. "On the Independence of the Axioms of Definiteness (Axiome der Bestimmtheit)", *Journal of Symbolic Logic*, 4, 1939: 69–72.
Russell, Bertrand. "Mathematical Logic as Based on the Theory of Types", in *Bertrand Russell: Logic and Knowledge; Essays 1901–1950*, R. Marsh, ed., London: George Allen and Unwin, 1956: 59–102.
Russell, Bertrand. "The Philosophy of Logical Atomism", in Marsh, 177–281.
Smullyan, Arthur F. "Modality and Descriptions", *Journal of Symbolic Logic*, **13**, 1948: 31–7. Reprinted in Linsky, Leonard, ed., *Reference and Modality*, Oxford: Oxford University Press, 1971, 35–43.
Tarski, Alfred. *Logic, Semantics, Metamathematics: Papers from 1923 to 1938*, translated into English and edited by Joseph H. Woodger. Oxford: Clarendon Press, 1956.
Whitehead, A.N., and Russell, B.A. ([PM]) *Principia Mathematica*, Cambridge: Cambridge University Press, 1910–1913, second ed., 1925–1927.
Weyl, Herman. *Das Kontinuum*, Leipzig: Veit, 1918.
Wolenski, Jan. *Logic and Philosophy in the Lvov-Warsaw School*, Dordrecht: Kluwer, 1989.

Chapter 15
Konstanty Michalski on Late Medieval Nominalism

Claude Panaccio

15.1 Konstanty Michalski (1879–1947)

Konstanty Michalski was one of the most influential medievalists in the field of philosophy during the first half of the twentieth century. In particular, he was with Pierre Duhem one of the great pioneers of the serious scholarly study of philosophy in the fourteenth century and he decisively contributed, through an incredibly patient labor, to the difficult disentanglement of the considerable stock of manuscripts that survive from this period. An indefatigable philologist with outstanding scholarship, he was the first to conduct an in-depth examination of an astounding quantity of manuscripts, which he located, identified, classified, attributed, dated, and sometimes summarized. A remarkable historical accomplishment indeed!

Michalski, however, was convinced that philosophy, not philology, should be the primary guide of historians of philosophy. One must 'protest', he says, 'against "historicism"', in so far as it is content with collecting more and more new material, thus adding to the 'already existing quantity of useless books'.[1] 'It is the historian of philosophy', he adds, 'and not the professional philologist, who must have the upper hand over the selection of texts'.[2] In other words, scholarly work on medieval texts should rest primarily on a good understanding of the philosophical problems at hand and be driven by a concern to outline what is really important philosophically in a given period. Actually, Michalski strived to bring to the fore, through a number of precise studies, a general picture not of the philosophical fourteenth century as a whole, but of a particular trend in it which he took to be of major significance and which he called 'the critical and skeptical trend'. His point was that the fourteenth century witnessed the progressive development of a deep distrust towards the cognitive capacities of human beings; and he stressed, in particular, the theoretical aspects

C. Panaccio (✉)
Université du Québec à Montréal, Montreal, QC, Canada
e-mail: panaccio.claude@uqam.ca

[1] Michalski (1969) 385. All the English translations from this book originally published in French are mine.
[2] Ibid.

S. Lapointe et al. (eds.), *The Golden Age of Polish Philosophy,* Logic, Epistemology, and the Unity of Science 16, DOI 10.1007/978-90-481-2401-5_16,

of the process, especially insisting on the major role played in it by the thought of William of Ockham.

What I aim to do here is to review some of the main Ockhamistic theses in which Michalski claimed to detect the germs of a new brand of skepticism, and examine in each case whether the thesis in question does indeed have the implications Michalski believed it had. I would thus like to offer a global assessment of Michalski's diagnosis about late medieval nominalism. My conclusion will be that despite many errors on particular points, and despite the fact that Michalski approached the fourteenth century from a standpoint which was very much akin on the whole to that of neoscholastic Thomism, his diagnosis turns out to be basically right.

Before I get to that, however, let me set out a little more precisely who Konstanty Michalski was.[3] He was born in 1879 and studied theology and philosophy in Louvain under Maurice De Wulf, graduating in 1911. He spent his whole career, after that, at the University of Krakow, where he started teaching as soon as 1914, being appointed in 1919 to one of the three chairs in philosophy that existed at that time in Krakow. He died in 1947. His interest for fourteenth century nominalism was first aroused by his early investigations into the history of philosophy in Poland, which was deeply influenced towards the end of the Middle Ages – especially in Krakow – by the nominalism of John Buridan (ca. 1295–1361). From there, Michalski got interested in the fourteenth century for itself and thus became one of the leading historians of medieval philosophy in Europe, Krakow being at the time, as Zenon Kałuża says, 'a real research center in medieval philosophy'.[4] From 1919 on, he was, notably, the secretary of the Research Commission on the History of Philosophy founded by the Polish Academy for Sciences and Literature. And he became in the 1920s the main promoter of an enormous international project consisting in putting together a vast 'Corpus of the Philosophers of the Middle Ages', only a part of which was finally accepted by the 'Union Académique Internationale'. Nonetheless, Michalski's initiative eventually gave rise to one of the most important endeavors of the twentieth philosophy in medieval philosophy, the celebrated *Aristoteles Latinus*.

In short, Michalski was a major figure of the 'Golden Age of Polish Philosophy'. According to Zenon Kałuża, he 'can be seen as the master of Salamucha and Bochenski since he presided their 'habilitations' and oriented their research in medieval philosophy'.[5] And he was, with Lukasiewicz, one of the initiator of the Krakow Circle, which aimed at 'reconciling Catholic thought with the precision of formal logic',[6] a group that existed from 1936 to 1939.

As to our knowledge of fourteenth century philosophy, Michalski's contribution is simply colossal. The core of it is to be found in six long papers written in French between 1921 and 1937 and collected in a single volume by the German medievalist

---

[3] For more on Michalski himself, see in particular Corvino (1959).

[4] Kałuża (1991) 99 (my translation).

[5] Ibid. 101.

[6] Ibid. 113–114.

Kurt Flasch in 1969 under the title *La philosophie au* XIV*e siècle. Six études*. Here are the titles of the papers (with my English translations):

- 'Les courants philosophiques à Oxford et à Paris pendant le XIVe siècle', (Philosophical trends in Oxford and Paris during the fourteenth century), 1921;
- 'Les sources du criticisme et du scepticisme dans la philosophie du XIVe siècle' (The Sources of Criticism and Skepticism in Fourteenth-Century Philosophy), 1924;
- 'Le criticisme et le scepticisme dans la philosophie du XIVe siècle' (Criticism and Skepticism in Fourteenth-Century Philosophy), 1926;
- 'Les courants critiques et sceptiques dans la philosophie du XIVe siècle' (Critical and Skeptical Trends in Fourteenth-Century Philosophy), 1927;
- 'La physique nouvelle et les différents courants philosophiques au XIVe siècle' (The New Physics and the Various Philosophical Trends in Fourteenth Century), 1928;
- 'Le problème de la volonté à Oxford et à Paris au XIVe siècle' (The Problem of the Will in Oxford and Paris in the Fourteenth Century), 1937.

In order to measure the erudition of these studies, one need only take a glance, for example, at the list of fourteenth-century authors to whom at least one paragraph and sometimes as much as a few pages is dedicated in the 1926 paper, 'Criticism and Skepticism in Fourteenth-Century Philosophy'. It includes William of Alnwick, William of Ockham, Adam Wodeham, John Rodington, Jean de Mirecourt, Robert Holkot, Roger Swineshead, Henry of Oyta, Thomas Wilton, Richard Fitzralph, John Baconthorp, John Buridan, Gregory of Rimini, Jean de Jandun, Henry of Harclay, Gerard of Bologna, Brinkley, Franciscus de Marchia, James of Eltville, Marsilius of Inghen, Hugues de Castro Novo, André de Castro Novo, John Duns Scotus, Hervaeus Natalis, Richard Campsall, Peter Aureol, Landulphus Caracciolo, Walter Chatton, Jean de Marchia, Bernard d'Arétie, Nicolas d'Autrécourt, Pierre d'Ailly, Albert of Saxony, John Gerson... and others! Michalski scrutinized in manuscript form important texts from each one of these authors. For the 1920s, this is astounding!

15.2 William of Ockham (ca. 1287–1347)

Michalski's central idea, as can easily be gathered from the mere titles of his studies, is that this extraordinarily rich intellectual fourteenth century can be characterized above all by the rise of what he calls criticism (a critical attitude, namely, towards the scope of human knowledge) and skepticism. And, as I said earlier, he assigns a central place to William of Ockham in this narrative. I would like to examine four theses he attributes to Ockham in this regard, and try to assess their skeptical import, if any. Three of these four theses are enumerated by Michalski in his 1921 paper and presented by him as both 'destructive' and highly influential in the fourteenth century.[7] They are the following:

(Th. 1) The thesis of the intuition of the non-existents;
(Th. 2) The nominalistic reinterpretation of simple supposition (*suppositio simplex*);
(Th. 3) The rejection of causality in its Aristotelian form.

To which I will add a fourth one that Michalski discusses in various places elsewhere in his writings, namely:

(Th.4) The idea that the objects of knowledge are propositions.

It will prove beneficial to review these four theses in the reverse order.

15.2.1 Propositions as the Objects of Knowledge

According to Ockham, the objects of knowledge – and belief, for that matter – are propositions, *mental* propositions more precisely, rather than external things or states of affairs.[8] This thesis, according to Zenon Kałuża, best characterizes nominalism in Michalski's view.[9] Kałuża bases this strong affirmation on a 1926 paper in Polish by Michalski, entitled 'The Rebirth of Nominalism', which, Kałuża says, 'was left unfinished and is unknown to historians'.[10] In Kałuża's rendering of this text (to which I had, personally, no independent access), 'the essence of nominalism consists in viewing judgment and its terms as the object of knowledge, and the assent to judgment as a psychological and intellectual act'.[11] Kałuża's claim is quite surprising and requires some discussion. For one thing, the thesis itself should certainly be rephrased, and the term 'judgment' replaced in it by 'proposition'. Judgment, for Ockham, is the act of assent itself, not its object.[12] Once this correction is made, it is undoubtedly true that the thesis in question was held by Ockham. What is surprising is Kałuża's claim that Michalski took it as expressing the *essence* of nominalism.

In the writings of Michalski with which I am familiar— the French papers—, he regularly applies the label 'nominalism' not only to Ockham but to John Buridan and his school as well, speaking frequently, for example, of 'Parisian nominalism' (Buridan taught in Paris, roughly, between the 1320s and the 1350s). But the thesis that mental propositions are the objects of knowledge, while it was defended by Ockham, was explicitly rejected by Buridan. And it was rejected, moreover, by various other influential nominalist thinkers of the fourteenth century, such as Adam

---

[7] Michalski (1969) 9–11.
[8] See for instance Ockham's Prologue to his *Commentary on Aristotle's Physics*.
[9] Kałuża (1991) 108.
[10] Ibid. 107 (my translation).
[11] Ibid. 108 (my translation).
[12] This is explained in details in the Prologue to his *Commentary on the Sentences*. See *Guillelmi de Ockham Opera Theologica* – henceforth: *OTh* – I, esp. 16–22.

Wodeham. This is something Michalski would certainly have known. The reduction of nominalism to the thesis according to which mental propositions are the objects of knowledge, consequently, does not correspond to Michalski's own use of the label in his published papers, which is much closer to the traditional – and, in my view, to the more interesting – sense according to which nominalism is a position about the so-called problem of universals: universals on this view are names, whether spoken or mental, rather than external things of their own.

This being said, it is true that mental propositions were taken by Ockham to be the objects of knowledge, and it is true that Michalski in his published papers correctly attributes the thesis to Ockham.[13] Whether or not the idea is essential to medieval nominalism (which seems wrong and, anyway, inconsistent with Michalski's own way of speaking), the question must still be raised: does it bring about skeptical consequences? Some believe it does. They reason that in taking propositions rather than external things to be the objects of knowledge, Ockham in effect was disconnecting knowledge from reality.

Yet, this is not so. Ockham always insists, when invoking this particular thesis, that although propositions are indeed what is known, these propositions, nevertheless, can very well be – and are, most of the time – about real external things. His point is that the component terms of known propositions, their subjects, for instance, may have – and do have, most of the time – what he calls 'personal supposition' (*suppositio personalis*): they can stand for real external things.[14] If I know, say, that horses are mammals, the object of my knowledge is the mental proposition 'horses are mammals', the main components of which are my concept of 'horse' and my concept of 'mammal'. But in order for the mental proposition 'horses are mammals' to be true – and therefore known – the concept of 'horse' must have personal supposition in it. It must stand there for real (singular) horses, and not for itself as a concept. My concept of horse, in Ockham's view, does represent real horses, and it does so, moreover, by virtue of its being a 'natural' mental sign of horses. The connection between knowledge and reality is thus not disrupted. Many recent commentators, on the contrary, actually insist that Ockham was a 'direct realist' in epistemology.[15] It is a mistake, consequently, to think that the thesis we are now considering confines human knowledge within the mental sphere.

But this is a mistake Michalski does *not* make. Whenever he attributes the thesis in question to Ockham, at least in the writings I am familiar with, he correctly points out that in normal cases the terms of the known propositions, for Ockham, stand for the things themselves, so that 'it in no way follows that Ockham was an idealist'.[16] In his 1921 paper in French, Michalski lists, as I said, three 'destructive' theses

[13] See e.g. Michalski (1969) 193–194.

[14] See for example Ockham's Prologue to his *Commentary on Aristotle's Physics*, *OPh* IV, 10–14, where he explains that the mental concepts which serve as subjects of the propositions of natural sciences normally stand in such propositions for real external things.

[15] See in particular Adams (1987) Chapter 13: 'Conceptual Empiricism and Direct Realism', 495–550.

[16] Michalski (1969) 194.

which, according to him, are to be found in Ockham (and to which I shall shortly return), but the reproach of idealism is not among them. Michalski, then, is entirely right about Ockham's defending the thesis according to which mental propositions are the objects of knowledge; but, in Ockham, the latter has no antirealist implications. I am only surprised that Michalski should have seen this position as the 'essence of nominalism', as Kałuża claims, but if he did, it was in an unpublished and unfinished paper which, I gather, should not be given too much importance.

15.2.2 Causality

I will be very short about the second thesis. Michalski holds that Ockham rejected the principle of causality as understood by Aristotle,[17] but his interpretation, here, is plainly mistaken. I will be content to refer the reader on this to Marilyn Adams's excellent and detailed account of Ockham's theory of causality in chapter 18 of her *William Ockham*.[18] The very first sentence of the chapter, actually, addresses the issue directly: 'Ockham's views about efficient causality', Adams writes, 'are among the most widely misunderstood parts of his philosophy';[19] this, I am afraid, squarely applies to Michalski's interpretation.

Michalski, however, does not develop the point very much. What he is really interested in is Ockham's criticism of the traditional proofs of the existence of God based on causality, the Thomistic proofs essentially. In Michalski's view, this criticism amounts to a rejection of a large part of traditional theology outside of the realm of rational philosophy.[20] And Michalski, on this latter point, is right: Ockham does challenge Aquinas's proofs and he is indeed much less optimistic than most of his medieval predecessors as to the scope of reason in theological matters. This Ockhamistic shift of attitude towards theology, however, does not rest on a rejection of the principle of causality 'in the form given to it by Aristotle', as Michalski claims.[21] Marilyn Adams, for one, explicitly raises the question: 'Does Ockham believe in necessary connections between cause and effect?'; to which she rightly and unambiguously replies: 'the answer is that he does',[22] an answer she then goes on to explain in details.

15.2.3 Simple Supposition

The Ockhamistic reinterpretation of simple supposition (*suppositio simplex*) – thesis (Th 2) above – raises a much more interesting issue for our present purposes. What medieval logicians called 'simple supposition' was characteristically exemplified

---

[17] Ibid. 11.
[18] Adams (1987) chap. 18: 'Efficient Causality', 741–798.
[19] Ibid. 741.
[20] Michalski (1969) 11.
[21] Ibid.
[22] Adams (1987) 754.

by a sentence such as the following, which everybody assumed to be true in some sense:

(1) Man is a species (*homo est species*).

What, it may be asked, does the term 'man' stand for in (1)? It clearly cannot stand for singular men as in:

(2) Man is a biped,

or

(3) A man was walking in the street.

In the latter two cases, 'man' was said to have 'personal supposition': it stands for the individuals which it applies to, or at least for some of them. But this cannot hold for (1): no singular human being is truly a species. 'Man', consequently, must have some special referential function in (1). This special function is what was called 'simple supposition' in the Middle Ages.

For most thirteenth-century logicians a term in simple supposition was taken to refer not to the external individuals it normally applies to, but to the corresponding common nature: 'man' in (1) was said to stand for human nature and the latter obviously entails an ontological commitment to universals. This interpretation of simple supposition, in other words, went along with a realist stance with respect to universals. Ockham, therefore, could not be happy with it, his most central philosophical tenet being that there is no such thing in external reality as a common nature or universal entity. This is precisely what his nominalism amounts to. Nevertheless, he had to provide some account for the truth of a sentence such as (1). His strategy in so doing was to stick to thirteenth-century terminology: he kept saying that 'man' has 'simple supposition' in sentence (1) – a vocabulary which John Buridan, for example, will simply drop shortly after—; but he radically reinterpreted the statement. A term taken in simple supposition cannot stand for a common nature in Ockham's view since there is no such thing! What he took it to stand for was the corresponding *concept*, understood as a mental sign: it is, literally, the concept 'man' which is a species, in Ockham's reinterpretation.

Simple supposition in this approach turns out to be something quite similar to what the Medievals called 'material supposition' (*suppositio materialis*), which is the use of a certain term to stand for itself as a spoken or written sign, as in:

(4) Montreal has eight letters,

where it is, normally, the written word 'Montreal' which is said to have eight letters, and not the city itself.[23] Simple supposition too, in Ockham's approach, is a special metalinguistic usage of a certain term, except that the referent in this case is not taken to be the spoken or written word, as in material supposition, but the mental

---

[23] On Ockham's theory of material supposition, see Panaccio and Perini-Santos (2004).

term or concept.[24] It is of this nominalistic reinterpretation of simple supposition, precisely, that Wilfrid Sellars once said that it was 'the first major breakthrough in the theory of categories' in the history of philosophy.[25] And it is this very same idea that Michalski, speaking from an entirely different point of view, described as 'the second undermining factor of the scholastic synthesis in general, metaphysics in particular, and even more particularly, of what is called the metaphysical degrees'.[26]

What then? Does Ockham's reinterpretation of simple supposition really have major skeptical implications? The answer to this ultimately depends on how we understand the connection between nominalism and skepticism. For the Ockhamistic reinterpretation of simple supposition is but a consequence in the end of Ockham's nominalistic rejection of universals as special extramental things Ockham in effect maintains the traditional idea that what a term taken in simple supposition stands for *is* a universal. Where he departs from his realist predecessors is in his nominalistic identification of universals with mental concepts. The question, then, comes down to deciding whether the nominalistic rejection of extramental universals implies, on its own, skeptical consequences with respect to the possibility of adequate knowledge.

This is a difficult problem, to be sure. But we certainly cannot take it for granted that nominalism automatically implies skepticism. This was not, at any rate, Ockham's own opinion. The *Venerabilis Inceptor* never thought that our general concepts are arbitrarily constructed and disconnected from reality. Of course, he did not see them as corresponding to general entities out there in the world, but nothing prevents them, according to his epistemology, from *correctly* classifying the external individuals which really exist independently of the human mind, all the more so, actually, since concepts, in Ockham's theory, are *naturally* formed in the human minds through causal processes. I like to quote the following revealing passage from his *Commentary on the Sentences*:

> It is not our intellect which makes a thing similar to another one, anymore than it makes Socrates white or Plato white.[27]

Socrates, Ockham says, is 'maximally similar' (*simillimus*) to Plato according to his very substance, and this is what allows the intellect to legitimately abstract a concept which is common to both Socrates and Plato, the concept of 'man' in this case.[28] In Ockham's own view, then, his nominalism in no way jeopardizes the adequacy of conceptual knowledge with reality.

---

[24] See William of Ockham, *Summa Logicae* I, 68 (*OPh* I, 207–208; Engl. transl. in Loux [1974] 198).

[25] Sellars (1970) 62.

[26] Michalski (1969) 10. See also 146: 'The interpretation of *supposition simplex* was the cause of the separation between the *moderni* or terminists in the narrow sense, and the *antiqui*'.

[27] William of Ockham, *Commentary in the Sentences* I, dist. 30, quest. 5, *OTh* IV, 385 (my translation).

[28] William of Ockham, *Commentary in the Sentences* I, dist. 2, quest. 6, *OTh* II, 211.

But Michalski never said it did. Contrary to many twentieth-century Thomist commentators, he rightly counted Ockham as a 'realist'.[29] Not a realist with respect to the external existence of universals of course, but a realist in epistemology, according to whom human knowledge does adequately represent extramental reality. What is 'dissolved' by the Ockhamistic reinterpretation of simple supposition in Michalski's view – remember the passage from *La philosophie au* XIV*e siècle* quoted just three paragraphs ago[30] – is:

- the scholastic synthesis in general,
- metaphysics in particular,
- and especially the idea of metaphysical degrees.

Of course, it is somewhat of an exaggeration to say that Ockham's nominalism dissolves metaphysics as a whole, in so far as it is itself a metaphysical position as has often been remarked.[31] Yet it is certainly true that it dismisses a certain hierarchical view of ontology, a certain idea of 'metaphysical degrees', since only singular entities are countenanced: Ockham's nominalism is a form of ontological egalitarianism. And it is true also that it repudiates a *certain* scholastic metaphysical synthesis, that of Aquinas namely (or of John Duns Scotus for that matter). If Michalski had in mind Thomistic metaphysics when he spoke of *the* scholastic synthesis – as can plausibly be supposed—, he is absolutely right to say that *this* synthesis was jeopardized by Ockham's reinterpretation of simple supposition – or by his nominalism, if you prefer.

15.2.4 The Intuition of Non Existents

In his 1921 paper, Michalski writes the following:

> But there is one thing that properly belongs to the Oxford's innovator [i.e. Ockham], it is the destructive idea which had a large influence on fourteenth century, and which we meet again and again in Ockham: God can produce an intuitive cognition in our mind without the object of this cognition being really present to our senses: man in this case is victim of an illusion, believing that the object really exists and does affect his sensitive organs.[32]

Nothing, in other words, could exclude the possibility of a deceitful God: I could be a brain in a vat! Ockham, Michalski explains, fully and repeatedly subscribes to the idea that God can cause directly whatever can be caused by natural things; and the consequences of it are simply pernicious:

---

[29] Michalski (1969) 238.
[30] See above n. 26.
[31] See, for example, the classical study of Philotheus Boehner (1947), precisely entitled 'The Metaphysics of William Ockham'.
[32] Michalski (1969) 9.

> The steady and ill-considered application of this principle in the field of knowledge was bound to engender distrust and skeptical spirit in philosophy of nature as well as in metaphysics and theology.[33]

What are we to think of this severe diagnosis?

Well, the first thing to say is that there is a major confusion here. Intuitive cognition, according to Ockham, is the cognition of a thing by virtue of which I can evidently *know* that the thing exists, if indeed it exists, or *that it does not exist, if indeed it does not exist*.[34] The intuitive cognition of a thing, in other words, is the simple apprehension of this thing which causes in us a *true* existential judgment: that the thing exists (if it does), or that it does not exist (if it does not). The intuition of non existents corresponds to the second case. This never happens in a purely natural way, Ockham thinks: we cannot naturally apprehend something which does not exist. But it is possible, nevertheless, by divine intervention. God, for example, could allow a prophet to intuitively grasp something which does not presently exist. The prophet in such a case would form the *true* judgment that the thing does not exist. Strictly speaking, Michalski, then, is wrong in thinking that the subject of such intuitions is 'victim of an illusion, believing that the object really exists' even though it does not exist. If an *intuitive* cognition is involved, the agent, on the contrary, will evidently know, in virtue of this intuitive cognition, that the thing does not exist. The Ockhamistic doctrine of the intuition of non-existents, in short, has by itself no skeptical implications, quite to the contrary.

This particular thesis, moreover, as Katherine Tachau has convincingly shown in the late eighties, had but a very small influence on fourteenth-century thought.[35] Contrary to what Michalski seemed to think, the thesis was generally rejected by Ockham's successors, even in the nominalist camp. Michalski, then, was twice mistaken, once with respect to his interpretation of Ockham, and once more with respect to the historical influence of the Ockhamistic doctrine of the intuition of non-existents.

This double mistake, however, does not prevent Michalski from being basically right. Ockham does explicitly admit that God can very well create in me the belief – and even the deep conviction – that a certain thing exists, even if it does not, or that it is present in front of me, even if it is not.[36] In such cases, as Elizabeth Karger has strongly stressed in a recent paper, the belief in question is not caused by an intuitive cognition, but directly by God.[37] No intuitive cognition is even involved in

---

[33] Ibid.

[34] See on this article 1 of the Prologue to Ockham's *Commentary on the Sentences*, *OTh* I, esp. 30–33; and his *Quodlibetal Questions* V, quest. 5: 'Do intuitive cognition and abstractive cognition differ from one another?', *OTh* IX, 495–500 (Engl. transl. in Freddoso and Kelley [1991] 413–417).

[35] See Tachau (1988).

[36] See William of Ockham, *Quodlibetal Questions* V, quest. 5, *OTh* IX, 498: 'Nonetheless, God can cause an act of believing through which I believe a thing to be present that is [in fact] absent' (Engl. transl: Freddoso and Kelley [1991] 416).

[37] See Karger (1999).

the process. And *a fortiori* no intuition of non-existents. This is the precise point Michalski is wrong about. Yet, it is still quite possible, according to Ockham's epistemology, that our experience in such a case of divine deceit would be subjectively indiscernible from the experience we would have through some real intuitive cognition of something existent. God, in other words, can insinuate in me the illusion that someone in particular is here in front of me even if she is not. It would not be a case of intuitive cognition, since intuitive cognition is always *by definition*, apt to cause *true* existential judgments. But subjectively, my experience would be indiscernable from the one I would have if the person in question was actually present.

Michalski, then, is right after all: Ockham does admit the possibility in principle of a deceitful God. And it is logically possible in his theory, consequently, that I should be in reality a brain in a vat! As far as I can see, in addition, it is also true that this logical possibility was generally accepted in the fourteenth century after Ockham.

15.3 Conclusion

What should we conclude as to Michalski's general diagnosis about criticism and skepticism in the fourteenth century, with regard, especially, to the connection between skepticism and nominalism? Well, first, we must pay attention to Michalski's own use of the term 'skepticism'. He does not use it, obviously, as we might be tempted to today, as a label for the philosophical position according to which it is impossible for human beings to know anything with a sufficient degree of certitude. Such radical skepticism is not to be found in fourteenth-century nominalism, and Michalski knows it: Ockham, Buridan and their successors all firmly believed in the possibility of science. When speaking of the rise of skepticism in fourteenth century, Michalski, it is true, does refer to a 'loss of confidence in the cognitive capacity of the human intellect',[38] but this 'loss of confidence', as he understands it, occurs mainly in the theological domain. It is first and foremost a loss of confidence in the capacity of human beings rationally to know the so-called theological 'truths'. Michalski's favorite example in this regard has to do with the alleged proofs of God's existence. Fourteenth-century nominalists – starting with Ockham – turned out to be much more skeptical than their thirteenth-century predecessors – Aquinas in particular – concerning the possibility of *demonstrating* in a purely rational and natural way the existence of a unique God, altogether omniscient and omnipotent. Michalski on this is quite right and his diagnosis is correct: there was indeed in the fourteenth century largely in connection with nominalism

[38] Michalski (1969) 38.

a general rise of skepticism concerning the possibility of natural theology, and a concomitant rise of fideism (an aspect on which Michalski also strongly insisted).[39]

Such was, in Michalski's eyes, the central component of what he called 'skepticism'. And he saw it as 'destructive' in so far as he himself subscribed – at least in his early years – to a neoscholastic approach of the Thomistic type in philosophy. Yet his general disapproval of the philosophical trends he was studying never prevented him from scrutinizing them with great scholarly rigor and intellectual respect, in striking contrast with what has often been the case with neoscholastic Thomists.[40]

It is true that theological skepticism, while crucially important for him, was not the only aspect of late medieval nominalism that bothered Michalski. He also repeatedly refers to the rise within natural philosophy of a distinctive epistemological attitude he calls 'probabilism' – or sometimes 'probabiliorism,'[41] which consists, with respect to a given problem, in holding that this problem cannot be demonstratively solved in an absolute way, but that a given answer to it, nevertheless, can be held as 'probable', or even as *more* probable than the rival ones. But in this too Michalski was right. This was indeed an attitude which became widespread during the fourteenth century. It was never entirely generalized – Ockham, for example, clearly thought that certain metaphysical and philosophical issues can be demonstratively decided, the problem of universals being perhaps the most prominent one – and never gave rise, consequently, to a radical form of skepticism. But it remains that on many questions Ockham, Buridan and the likes declared themselves ready to settle for merely 'probable' answers. A good example of this in Ockham is his position with respect to the ontological status of mental concepts: are they purely ideal items with a special mode of existence of their own or should they be identified with mental acts? After having favored the former position in his earlier works, Ockham finally settled for the latter as more economical since it does not multiply beyond necessity the types of entities acknowledged in the ontology. But he never took this 'Razor' argument to be decisively demonstrative: the theory that concepts are mental acts was merely taken by him to be more 'probable' than the ideal object theory.[42]

This is, then, another manifestation of the increasing reserve on the part of fourteenth-century philosophers – nominalists in particular – as regards the ability of human reason for decisively adjudicating between rival theoretical possibilities. And

---

[39] It must be noted, however, that Michalski wrongly attributes to Ockham – with most medievalists of the early twentieth century – a treatise called the *Centiloquium* (*OPh* VII, 371–505), which goes quite far in the way of theological skepticism, but which is not considered anymore as an authentic Ockham writing (see in particular Boehner [1944]). This being said, it remains true that Ockham severely criticized the Thomistic proofs of God's existence, and that Buridan, for instance, thought that the immateriality and immortality of the human soul could not be rationally proved.

[40] Let me stress in addition that such derogatory terms as 'destructive idea' appeared mostly in the 1921 paper, and much more rarely in Michalski's subsequent writings.

[41] See, for example, Michalski (1969) 177.

[42] See on this Panaccio (2004), esp. Chapter 2, Section 2, 23–27.

one which clearly lies outside the field of theology. Michalski's general diagnosis, after all, turns out to be right in this respect as well. Even more radically, Michalski, as we saw, was also globally right – even though he was mistaken in the details – in thinking that Ockham and many of his successors did admit the logical possibility of divine deceit with respect to our apparent grasping of external objects. And this is also, of course, a certain form of skepticism.

My conclusion, in short, is that Michalski's general diagnosis about fourteenth-century philosophy was basically correct. My disagreement consists in being more inclined than he was to take the forms of theological skepticism and epistemological probabilism which became widespread in late medieval nominalism to be philosophically sound. That the thirteenth-century synthesis and the realistic acceptation of universals that went along with it were repudiated can be seen, I gather, as a benefit rather than a loss. And above all, the development of a less dogmatic, less naïve attitude towards human knowledge was, I take it, a major step in the history of epistemology. William of Ockham's position in this respect appears especially interesting. On the one hand, he does admit the logical possibility of radical illusion. But he also acknowledges on the other hand that we never have any good reason to think that we are in fact radically deceived, and he considers human beings to be *de facto* equipped with a generally reliable cognitive apparatus rooted in the causal components of intuitive grasping. Such a reliabilist and causal approach still seems to be a highly commendable course in epistemology. While Michalski's diagnosis is correct, then, his mistake, as I see it, is to disapprove of these developments he so keenly studied. Yet it must be added that his disapproval does not appear to be that strong, in the end: Michalski was much more interested in understanding fourteenth-century thought than he was repelled by it.

Bibliography

Adams, Marilyn McCord (1987), *William of Ockham*, South Bend, University of Notre Dame Press.

Boehner, Philotheus (1944), 'The Medieval Crisis of Logic and the Author of the Centiloquium Attributed to Ockham', *Franciscan Studies* 4, 151–170; reprinted in Boehner, 1958, 351–371.

Boehner, Philotheus (1947), 'The Metaphysics of William Ockham', *The Review of Metaphysics* 1/4 (1947–1948), 59–86; reprinted in Boehner (1958), 373–399.

Boehner, Philotheus (1958), *Collected Articles on Ockham*. St. Bonaventure, The Franciscan Institute.

Corvino, Francesco (1959), 'Konstanty Michalski (1879–1947)', *Rivista critica di storia della filosofia*, 1, 206–220.

Freddoso, Alfred J., and Francis E. Kelley (1991) *William of Ockham. Quodlibetal Questions*, Eng. Trans. New Haven, Yale University Press.

Kałuża, Zenon (1991), 'Pologne: Un siècle de réflexions sur la philosophie médiévale', in Imbach and Maierù, 97–130.

Karger, Elizabeth (1999), 'Ockham's Misunderstood Theory of Intuitive and Abstractive Cognition', in Spade, 204–226.

Imbach, Ruedi and Alfonso Maierù (eds.) (1991), *Gli studi di filosofia medievale fra otto e novecento*, Rome, Edizioni di storia e litteratura.

Loux, Michael J (1974), *Ockham's Theory of Terms. Part I of the Summa Logicae*, South Bend, University of Notre Dame Press.
Michalski, Konstanty (1969), *La philosophie au* XIV*e siècle. Six études*. Kurt Flasch (ed.), Frankfurt, Minerva.
Panaccio, Claude (2004), *Ockham on Concepts*. Aldershot, Ashgate.
Panaccio, Claude, and Ernesto Perini-Santos (2004) 'Guillaume d'Ockham et la *suppositio materialis*', *Vivarium* 42, 202–224.
Sellars, Wilfrid (1970) 'Toward a Theory of the Categories.' In Foster and Swanson, 55–78.
Tachau, Katherine H. (1988), *Vision and Certitude in the Age of Ockham*, Leiden, Brill.
William of Ockham, *Opera Philosophica*. 7 vols., Gedeon Gál et al. (eds.), St. Bonaventure, The Franciscan Institute, 1974–1988.
William of Ockham, *Opera Theologica*, 10 vols., Gedeon Gál et al. (eds.) St. Bonaventure, The Franciscan Institute, 1967–1986.

Chapter 16
Jan Salamucha's Analytical Thomism

Roger Pouivet

Quod intelligimus debemus rationi, quod credimus auctoritati[1]

16.1 Salamucha and Analytical Metaphysics

One may have interest in Polish philosophy out of Polonophilia or because one aims at restoring Polish philosophy's right to historical recognition, and both are excellent reasons. In my opinion, Polish philosophy anticipated what constitutes the best part of analytical philosophy *today*. This is the case, in particular, of Jan Salamucha, a philosopher whose renown is lesser than that of Tarski, Łukasiewicz or Kotarbiński, but whose work has much in common with metaphysics and the philosophy of religion in their current analytical version.

Explaining how Salamucha anticipates today's analytical philosophy requires some historiographic background. The canon as far as the history of analytical philosophy is concerned consists in claiming that it was born of the rejection of metaphysics. It is supposed to have progressively eradicated it, first through logical analysis (Wittgenstein, Russell, Carnap), then through the analysis of ordinary language (Ryle, Austin), and finally by bestowing an increasing importance to what is "relative" in knowledge and by paying attention to its social context. Among the many reasons one may have to contest this tale, two stand out in relation to Salamucha's place within 20th century philosophy.

First, one has come to realise that phenomenology and analytical philosophy have a common origin in Austrian philosophy at the turn of 19th and 20th centuries.[2] Brentano, Meinong, the first Husserl (sometimes called the "Austrian" Husserl), were engaged in investigations that led them to question the concepts of identity, of substance, of property, of relations, of truth, etc. Their theories share neither the form nor the method of German idealism and they modestly, and in close relation with the sciences of the day, try to understand the nature of objects, properties, relations, etc. In this respect, their theories can be said to be substantially determined

R. Pouivet (✉)
Université de Nancy ii, Archives Poincaré (CNRS), Nancy, France

[1] Saint Augustine, quoted by Salamucha in his article "Wiara" ([1936] 1997, 112; 2003, 274).

[2] For example, Simons, 1992; Smith, 1994.

S. Lapointe et al. (eds.), *The Golden Age of Polish Philosophy,* Logic, Epistemology, and the Unity of Science 16, DOI 10.1007/978-90-481-2401-5_17,

by metaphysical concerns. Russell, Arthur Prior, Roderick Chisholm and Gustav Bergman are as many missing links between these Austrian philosophers and an important part of today's analytical philosophers who are explicitly "metaphysical", such as David Armstrong, David Lewis, Peter Van Inwagen, Peter Simons, E. J. Lowe, etc. Consequently, the analysis of everyday language, the cult of the ordinary supposedly derived from the second Wittgenstein – I say "supposedly", because I think there are other possible interpretations of the second Wittgenstein[3] – are in no way the necessary upshot of the failure of logical positivism. On the contrary, in many respects, Austin and the neo-Wittgensteinian represent a mere *parenthesis* between two ways of dealing with metaphysical questions. The first, at the beginning of the 20th century, with the first Wittgenstein, Moore, Russell; the second, today. "The Elimination of Metaphysics through Logical Analysis of Language" of Carnap, amounted to the firm rejection of a type of philosophical literature, namely Heidegger's. Yet Carnap's *Aufbau*, as even a superficial reading will show, is a metaphysical book.

The second reason why the standard narrative of the development of analytical philosophy should be put into question is that it leaves no place for the main proponents of contemporary metaphysics, David Armstrong, Roderick Chisholm, David Lewis, Peter Van Inwagen, Peter Simons, E. J. Lowe, Joshua Hoffman and Gary Rosenkrantz, and seems to ignore the contemporary development of analytical metaphysics in the United States, Australia, New Zealand, and even in the United Kingdom. Where ought one to situate the metaphysical preoccupations of semanticists concerned with modalities, such as Saul Kripke or Alvin Plantinga and, in particular, their use of the notion of possible worlds? If the accepted tale were correct, these philosophers would have to be the outcome of spontaneous generation or the unfortunate offshoots of an old metaphysical root, which the defoliants of the logical analysis of language failed to eradicate. But, these metaphysicians are parts and parcels of analytical philosophy and we find among them some of the best logicians. The close readers of Carnap, Quine, or Strawson, are reconsidering the more deliberately metaphysical sides of the works of Russell, Wittgenstein and Moore.

Hence there is another tale to be told. And it is. The *Oxford Handbook of Metaphysics* (Loux and Zimmerman Dean 2003)[4] – and the introduction in particular – is a hymn to the renaissance of metaphysics. It looks back to the themes and methods which belonged to the grandfathers and fathers of analytical philosophy as well as to other underestimated figures of Anglo-Saxon philosophy of the late 19th and at the beginning of the 20th centuries, such as McTaggart or Prior. This new tale needs, however, some adjustment as regards two distinct but intellectually close philosophical movements. The first was developed in Great-Britain in the 1950s around Elizabeth Anscombe and Peter Geach (very nearly Polish ! though not our

[3] Pouivet, 1997.
[4] Also Zimmerman, 2004; Nef, 2004.

problem here). The second concerns Polish philosophy in the interwar period, the Lvov-Warsaw School in particular. For one thing, Łukasiewicz's metaphysical concern for the question of determinism, for instance takes the form of a reflection on the principle of contradiction[5] and saw the birth of three-valued logic.[6] Kotarbiński's reism seeks to answer the question of the nature of reality's fundamentals – an "ontological deflationist" is certainly no less a metaphysician. The great figures of the Lvov-Warsaw School may be considered to pre-empt the turn to logic in the treatment of fundamental metaphysical questions – which is the real point of analytical metaphysics. And this, too, is true of Salamucha.

16.2 Jan Salamucha and the Cracow Circle

Jan Salamucha was a Catholic priest whose philosophical education took place at Warsaw University, from 1924 onward. He studied with Kotarbiński and Leśniewski. In 1926, Salamucha was granted the title of *magister theologiæ speciæm philosophiæ* for a study on Aristotle's *Categories*, and in 1927 he became a doctor in Christian philosophy – a fact I will investigate later – on the basis of a thesis on *The Theory of Modal Consequence in Aristotle*. He then joined the Gregorian university (Rome) and wrote *De deductione apud Aristotelem et S. Thomam,* pursuing his habilitation at the Jagellonian University. He was supervised by the medieval specialist Konstanty Michalski[7]. In Bocheński, Sobociński and Drewnowski he founds, within the Lvov-Warsaw School, a group of explicitly Catholic philosophers and the latter who would later come to be known as the "Cracow Circle". Salamucha was deported in 1939 to Sachsenhausen, then Dachau, and was freed after the intervention of Heinrich Scholz, the historian of logic. Salamucha was assassinated in 1944, during the Warsaw Uprising by troops belonging to the Russian national army of liberation who were fighting on the German side.

Salamucha was a Thomist. According to John Haldane, analytical Thomism aims at "applying the methods and ideas of twentieth century philosophy – prevailing in the English speaking area – in the context of the vast network of ideas initiated and developed by saint Thomas" (1997, 486).[8] John Haldane speaks about British philosophers like Peter Geach, Elizabeth Anscombe, and himself. But before it became a British affair, Analytical Thomism was invented by the

---

[5] See Łukasiewicz, 2000.

[6] It is interesting to note that the title of the edition by J. Jadacki of a collection of articles by Łukasiewicz is *Logika i metafizyka* (1998). Concerning which point, Łukasiewicz, affirmed that "it is enough to state that logistics does not include, either explicitly or implicitly, a determinate philosophical doctrine, neither does it surreptitiously promote any antireligious propensity" (1970, 234)

[7] On Michalski, see Claude Panaccio's paper in the present volume.

[8] See also, Haldane, 2004.

Polish philosophers of the Cracow Circle. That the works of these Polish analytical Thomists should be so little known is probably due to their having written almost uniquely in Polish, a notoriously difficult language, and to the fact that the iron curtain not only seems to have separated Central Europe from Western Europe, but also induced a sort of cultural amnesia in the West.

Jan Salamucha found in the logical and semiotic instrumentation of the Lvov-Warsaw School a privileged means to support the fundamental theses of Thomism.[9] What's more, the analytical method of the Lvov-Warsaw School allow him to revive certain interesting aspects of Scholastic philosophy, which, under the very influence of the modern philosophy which they meant to contest, the neo-schoolmen had left aside: the logic of consequences, the theory of suppositions, the theory of antinomies, modal logic. As Bocheński put it, "neoscholastic philosophers were somehow too progressive" (1989, 13). For the most part, they seemed to consider that modern philosophy represented a progress or, at least, that one should take into account what it has "brought" us. And yet, they might be said not to have been progressive enough, ignorant as they were of Frege's and Russell's logic. One may conclude that the neoschoolmen were mistaken as to the true nature of intellectual progress. As Bocheński observes, to write on the Trinity, knowing that there is a theory of ternary relations might come in handy; to write on the infinite series implied by the proofs of the existence of God, it is important to be conversant with the theory of serial relations as worked out by the logicians. Borrowing as it did from the contents and methods of the recently rediscovered Aristotle, the *Summa Theologiae* of Medieval philosophers had achieved, in their time, the best in philosophy. Why then refuse what is best in current thought, asked the philosophers of the Lvov-Warsaw School (Drewnowski, 1937, 223) and which is embodied by analytical philosophy in its two contemporary forms: Russell's logical analysis and the Lvov-Warsaw School.

The Cracow Circle is sometimes said to have been characterised by the use of formal methods in matters of rational theology. Salamucha's philosophy cannot be reduced, though, to his famous formalisation of the *ex motu* proof in the *Summa contra Gentiles,* too often claimed to be his main achievement. The publication in 1997 of a volume entitled *Wiedza i Wiara* (Knowledge and Faith), gathering the greatest part of his philosophical writings, makes it possible to rectify this overly narrow perspective.[10] It now emerges that his works belong to three genres:

1. He realised historical investigations on medieval philosophy which both answered the necessity of historical investigations (the use of original texts) and were *au fait* with what was being done at the time in logic and the philosophy of logic. Salamucha's method obviously anticipated the contemporary approaches of American medieval studies, for instance, the one recommended

[9] See Pouivet, 1999, 2000, 2001, 2003.

[10] One also finds an English volume of articles by Salamucha (2003).

by Norman Kretzmann (1982) in the introduction to the *History of Late Medieval Philosophy*.[11]

2. He applied formal methods to the *ex motu* proof or to other aspects of scholastic medieval philosophy as found in the works of Aquinas, Ockham, Duns Scot and others. The idea was to put the validity of the arguments supported by these philosophers to the test of contemporary logic.
3. He published articles in the philosophy of religion and ethics. The latter are manifestly aimed at a wider public, beyond specialised historians of medieval philosophy, or philosophers of logic. In my opinion, these articles epitomize very closely the idea of analytical Thomism.

I wish to document this third point, and in order to do so I will examine some of Salamucha's ideas from two articles, "Tomizm jako *philosophia perennis*" (Thomism as *philosophia perennis*)[12] and "Teologia i filozofia" (Theology and philosophy), published in 1946 in *Tygodnik Powszechny*, a Catholic publication.[13]

16.3 Ens et Verum Convertuntur

In the last part of the paper entitled "Thomism as *philosophia perennis*", Salamucha proposes a classification of philosophical systems on the basis of a certain way of conceiving of the relation between the domain of a philosophical system and the method it favours. Now, as regards domain and method, Salamucha distinguishes between maximalism and minimalism. Domain-maximalism is a function of the range of objects relevant to a system. A domain is at its maximum when it includes reality in its totality. From the methodological standpoint, maximalism consists in saying: "only theses verified by means of experimental methods can be said to be scientific, or theses integrated in a deductive system organised on the basis of strictly logical methods according to a clearly formulated protocol" (1997, 62). Methodological minimalism, roughly, corresponds to the intuitive approach Salamucha associates with Bergson: "the source of philosophical knowledge is an intuition ideally purified of all rational and experimental elements, it is intuition

---

[11] "By combining the highest standards of medieval scholarship with a respect for the insights of contemporary philosophers, particularly those working in the analytic tradition, we hope to have presentend medieval philosophy in a way that will help to end the era during which it has been studied in a philosophical ghetto, with many of the major students of medieval philosophy unfamiliar and un sympathic with twentieth-century philosophy, and with most contemporary work in philosophy carried out in total ignorance of the achievements of the medievals on the same topics" (Kretzmann, 1982, 3). C. Panaccio's research might be taken to embody today what Kretzmann looked forward to (1991).

[12] This paper is translated into French (by D. Sikora and R. Pouivet), in *Revue des Sciences Religieuses,* note 1, Janvier 2006.

[13] Czesław Miłosz pretended that *Tygodnik Powszechny* was for forty years the only representative of free thought from the Elbe to Vladivostok! These two articles are not to be found in the collection of texts by Salamucha translated into English (2003), but in *Wiedza i Wiara*, published in 1997.

as understood through the analogy with artistic inspiration and religious ecstasy" (1997, 62). (I leave aside the problem of whether this interpretation of Bergson is correct or not.) Domain-minimalism, which is illustrated by 19th century French positivism and the Vienna Circle, restricts "the range of scientific research only to phenomena confirmed by experience, and to *a priori* (tautological) constructions, as in the case of mathematics and of logic" (1997, 62). (One may question the validity of the attribution of a minimal domain to Comte's system, but my aim here is not to evaluate this interpretation.)

Given what precedes, one may according to Salamucha conceive of four types of philosophical systems:

	Domain (range)	Method
(1)	Maximal	Maximal
(2)	Maximal	Minimal
(3)	Minimal	Maximal
(4)	Minimal	Minimal

According to Salamucha, Bergson's system is an instance of (2). 19th century positivism and the Vienna Circle are examples of (3). I will come back on (1) in more detail in what follows. Salamucha did not propose examples of (4) which, according to him lacks interest: the method of a philosophical system of this type would remain uncertain and its cognitive benefit would apparently be very limited. Salamucha makes it clear that "'neat' classifications are only to be found in mathematics and logic".

Methodological maximalism and domain minimalism go together since the high level of methodological requirement makes the application of the system to certain kinds of objects difficult. Methodological minimalism and domain maximalism go together since the low level of methodological requirements allows for an application to indefinitely many objects, if not all. Methodological maximalism is consequently a principle that limits the scientific field to a certain type of objects for which a scientific approach is appropriate.

What kind of systems would be instances of (1)? Salamucha considers that, "the union between maximalism of extension and maximalism of method would represent a sort of paradox. History, however, has already been found to offer such paradoxical systems, with Aristotle, Descartes, Leibniz, Kant, and most obviously, Aquinas" (1997, 62). For Salamucha,

> Thomism admits of no limitation in the field of scientific research. Ontological truth is its motto: *ens et verum convertuntur* – all that exists is God or comes from God, that is why

> all that exists is potentially liable to human knowledge. Thomism makes philosophy impervious to insufficiently motivated theses: each philosophical thesis should be motivated, as strictly as possible, even if this motivation is specific. Its truth-value depends on the nature of its motivation. This must be as strict as possible, but specific, Thomism advocating no single universal method. Methods depend on the field of research; a method is subordinate to the object. Roughly speaking, one could say that a Thomist considers that all can be examined scientifically, whereas impressions and feelings do not foster any scientific knowledge. One is not allowed to pretend to solve a problem in matters of sciences. One must systematically delve into things, all things, adapting and inventing tools of knowledge. (1997, 63)

The consequence is that Thomism unites methodological maximalism and domain maximalism. But what makes it different from systems of type (1), though? It is the subordination of the method to the objects, that is to say, realism: the world and what it truly contains is mind-independent (at least human-mind independent). But one may of course find realist systems other than Thomism. In the case of Thomism, the specific difference is theological. Thomism considers theology to be the negative norm of philosophy.

16.4 What is Christian Philosophy?

The recourse by Thomism (even though analytical), to a negative theological norm naturally leads to an objection which can be formulated as follows: Salamucha, in his defence of clarity, of argumentative rigour and of scientific earnestness, emphasises methodological maximalism. But by resorting to a negative theological norm, is what he achieves anything more than a catechism – a strict catechism, and, for all that, absolutely foreign to philosophy? What would happen if philosophy led to theses contradicting the Christian dogmas? Would it then have to choose?

Salamucha answers this objection in the article entitled "Theology and Philosophy". He first proposes the following figure:

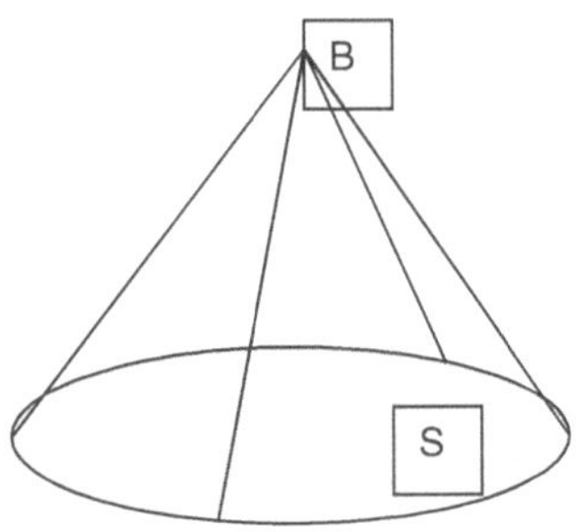

B stands for God, and S for the empirical world. The lines joining B and S stand for the relation between God and the world. Specific sciences are limited to the study of the elements and relations in S. The object of theology is B and the relation between B and S. The object of philosophy is the entire cone. Consequently, "all the conflicts between theology and specific sciences are nothing but appearances, for there exists no common ground on which these much anticipated conflicts might occur" (1997, 48).

Salamucha then proposes the following diagram:

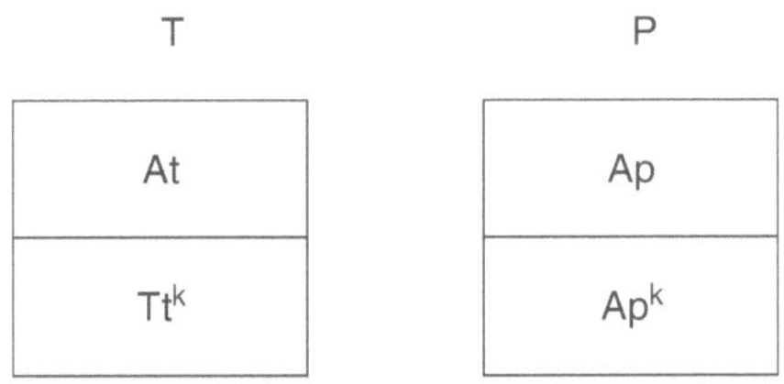

T stands for theology; *P* stands for philosophy. According to Salamucha, the first level refers to deductive systems. "At" stands for the axioms of theology; "Ap" for the axioms of philosophy. "Tt^k" and "Ap^k" stand respectively for the theses of theology and philosophy that are derived from the axioms, that is, for theorems – the index "k" indicates the level of the latter. In so far as the two axiomatic domains are distinct, a conflict between theology and philosophy could only be apparent. But what if "Tt^k" is "God exists" and "Ap^k", "God does not exist"? One could try to argue the theses in question belong to distinct axiomatic domains, but since philosophy is represented by the cone as a whole, the answer would be of no avail. Salamucha explains:

> In the case of an explicit contradiction between a theological and a philosophical thesis, one is faced with two logically acceptable possibilities: (1) the theological thesis is founded and the philosophical thesis is unfounded, and must be eliminated, or conversely (2) the philosophical thesis is founded and the theological thesis unfounded. He who chooses the second possibility immediately ceases being a Catholic since he rejects the truth revealed to us. A Catholic has no choice but (1); the thesis which contradicts a theological thesis, whichever, must be eliminated from the philosophical system. It is clear that after the elimination, for the reasons just stated, of thesis Tp^k from the philosophical system, a Catholic is not allowed to replace it in his philosophical system, by a theological thesis Tt^k. If he did so, he would transgress the methodological differences between the specificities of theology and philosophy. The elimination of thesis Tp^k from the philosophical system leaves a gap in this system which must be made up for by means of philosophical methods. The motto according to which theology is a negative form of philosophy expresses the fact that one depends on the other. (1997, 49-50)

Salamucha's analytical Thomism thus turns theology, as an axiomatic of Revelation, into the negative norm of philosophy. But he also ascribes to philosophy the obligation to account through the philosophical means that pertain to its own axiomatic, and which are not founded in Revelation, for certain theses negatively imposed upon it. If a philosophical thesis has to be negated, it must be replaced by another one "conforming to the philosophical method". In this respect, Salamucha's analytical Thomism appears to take up the traditional program of rational theology we find today, for instance, in the works of Richard Swinburne.[14]

And yet, is this nothing more than a reassertion of the old slogan making philosophy the servant of religion? Is not Christian "philosophy" an absurdity? Is it not the case that the word "Christian", because it entails the respect of certain revealed dogmas, contradicts the very project of philosophy as free and rational thought! How can a philosopher discuss the truth of non philosophical theses, and especially theological ones, serving as negative norms? Salamucha's counter-objection is to be found in his tacit criticism of Heidegger's observation that Christian philosophy is like a square circle. The idea of Christian mathematics or of Christian physics is absurd because there would be no difference between Christian mathematics or Christian physics, and non-Christian mathematics or non-Christian physics.

On the other hand, Christian philosophy may be distinguished from a philosophy which takes no account of Christian theology, and examples of it can be found in history. It is true that if philosophy kept to strict and infallible norms, theology would have no place in it, even as a negative norm, because such methods would protect it from errors. But philosophy is far from having established such strict, infallible and satisfactory methods yet. Philosophy is still a risky matter concerning which theology is often beneficial to the Catholic. (1997, 50)

How far does this argument go? It does show that the expression "Christian philosophy" is not absurd in the way in which the expression "Christian physics" would be. But does it not also confirm our doubts concerning the unsound dependence of philosophy on religious dogmas?

It is a pity that Salamucha did not think of comparing the situation of philosophy in its relation to theology with its situation in relation to sciences. For it comes, or ought to come, as no shock to anyone that a philosopher should doubt a philosophical thesis when it contradicts an accepted scientific theory. For a Christian philosopher, theological dogmas constitute a negative norm, in precisely the same way that for some philosophers, scientific explanations constitute a negative norm to philosophy itself. The *analytical Thomism* of Salamucha consists in philosophising while accepting theological theses. Philosophy cannot make use of such theses though, but uses them as negative norms, just as one may use certain scientific theses as negative norms in the field of philosophy.

Anthony Kenny observes that in *Principia Mathematica,* Russell needs hundreds of pages to prove the proposition "$2 + 2 = 4$" (1993, 12). This proposition is a negative norm in the system of *Principia Mathematica.* While presupposing the truth of this proposition, Russell intends to prove it on the basis of the means inherent in the system of *Principia.* A failure would not have led either him or his co-author, Whitehead, or anybody else, to give up thinking that $2 + 2 = 4$. With Salamucha, analytical Thomism adopts a similar attitude concerning a proposition like "God exists" or "God is absolutely good". This is what makes it a Christian philosophy.

[14] Also Swinburne, 1979; Clavier, 2004.

16.5 Conclusion

Salamucha foreshadows some aspects of what has become part of contemporary analytical philosophy. More particularly, he develops, on the basis of the method of the Lvov-Warsaw School, a version of analytical Thomism which is in no way limited to an attempted formalisation of the *ex motu* proof, but includes the general question of the relation between theology, religion and philosophy, as shown above, and ethical problems which I was unable to tackle here. Salamucha, who was one of the 200 000 victims of the Warsaw Uprising, was too young – forty one – at the time of his death to have fully developed his ideas. Nonetheless, he deserves our full philosophical attention.

Bibliography

Bocheński, Józef (1989), 'The Cracow Circle', K. Szaniawski (ed.), *The Vienna Circle and the Lvov-Warsaw School*, Dordrecht, Kluwer.

Clavier, Paul (2004), *Qu'est-ce que la théologie naturelle?* Paris, Vrin.

Drewnowski, Jan (1937), 'Neoscholasticism and the Demands of Modern Science'. *Axiomathes*, vol. 2, 1993.

Haldane, John (1997), 'Analytical Thomism: A Brief Introduction', *The Monist*, vol. 80/4, 485–486.

Haldane, John (2004), *Faithful Reason, Essays Catholic and Philosophical*, London, Routledge.

Kenny, Anthony (1993) *Aquinas on Mind*, London, Routledge.

Kretzmann, Norman (1982), 'Introduction', N. Kretzmann, A. Kenny, J. Pinborg, E. Stump (eds.), *The Cambridge History of Late Medieval Philosophy*, Cambridge, Cambridge University Press.

Loux Michael J., and Zimmerman Dean W. (2003), *The Oxford Handbook of Metaphysics*, Oxford, Oxford University Press.

Łukasiewicz, Jan (1970), *Selected Works*, Amsterdam, North-Holland.

Łukasiewicz, Jan (1998), *Logika i Metafizyka* (Logic and Metaphysics), Warsaw, Towarzystwo Naukowe Warszawskie.

Łukasiewicz, Jan (2000), *Du principe de contradiction chez Aristote*. French tr. D. Sikora, Paris, Édition de l'éclat [*O zasadzie sprzeczności u Arystotelesa*, 1910].

Nef, Frédéric (2004), *Qu'est-ce que la métaphysique?*, Paris, Gallimard.

Panaccio, Claude (1991), *Les mots, les concepts et les choses: La sémantique de Guillaume d'Occam et le nominalisme d'aujourd'hui*. Montréal/Paris, Bellarmin/Vrin.

Pouivet, Roger (1997), *Après Wittgenstein, saint Thomas*. Paris, Presses Universitaires de France.

Pouivet, Roger (1999), 'Łukasiewicz: de l'aristotélisme autrichien à l'aristotélisme polonais', *Philosophiques*, 26/2, 263–277.

Pouivet, Roger 2000, 'Logique et éthique: la nature du principe de contradiction chez Łukasiewicz', preface to Łukasiewicz, 2000, 7–34.

Pouivet, Roger 2001, 'Faith, Reason, and Logic', T.L. Smith (ed.), *Faith and Reason*. South Bend, in St Augustine's Press, 277–282.

Pouivet, Roger 2003, 'Le thomisme analytique, à Cracovie et ailleurs.' *Revue Internationale de Philosophie*, n°3 (Analytic Philosophy of Religion).

Salamucha, Jan (1997), *Wiedza i Wiara., pod redakcją J. Jadackiego i K. Świętorzeckiej*, Lublin, Towarzystwo Naukowe Uniwersytetu Lubelskiego.

Salamucha, Jan (2003), *Knowledge and Faith*, K. Świętorzecka and J. Jadacki (eds.), Amsterdam, Rodopi.

Smith, Barry (1994), *Austrian Philosophy, The Legacy of Franz Brentano*, Open Court, Chicago.

Simons, Peter (1992), *Philosophy and Logic in Central Europe from Bolzano to Tarski*, Dordrecht, Kluwer.
Swinburne, Richard (1979), *The Existence of God*, Clarendon Press, Oxford.
Zimmerman, Dean W. (2004), 'Prologue: Metaphysics After the Twentieth Century.' *Oxford Studies in Metaphysics,* vol. 1, ix–xvii.

Index

FSC
www.fsc.org
MIX
Papier | Fördert
gute Waldnutzung
FSC® C083411

Kolibri 360 GmbH
Ferdinand-Jühlke-Straße 7
99095 Erfurt, Deutschland
produktsicherheit@kolibri360.de